MEMOIRS
of the
American Mathematical Society

ISSN 0065-9266 (print)
ISSN 1947-6221 (online)

Number 1566

On Infinite Periodic Band Matrices and the Periodic Toda Flow

Luen-Chau Li

June 2025 • Volume 310 • Number 1566 (first of 6 numbers)

AMERICAN MATHEMATICAL SOCIETY

Library of Congress Cataloging-in-Publication Data

Cataloging-in-Publication Data has been applied for by the AMS.
See http://www.loc.gov/publish/cip/.
DOI: https://doi.org/10.1090/memo/1566

Memoirs of the American Mathematical Society

This journal is devoted entirely to research in pure and applied mathematics.

Subscription information. Beginning in 2024, Memoirs will be published monthly through 2026. Memoirs is also accessible from www.ams.org/journals. The 2025 subscription begins with volume 305 and consists of twelve mailings, each containing one or more numbers. Individual subscription prices for 2025 are as follows. For electronic only: US$1044. For paper delivery: US$1197. Add US$65 for delivery within the United States; US$200 for surface delivery outside the United States. Upon request, subscribers to paper delivery of this journal are also entitled to receive electronic delivery. For information on institutional pricing, please visit `https://www.ams.org/publications/journals/subscriberinfo`. Subscription renewals are subject to late fees. See `www.ams.org/journal-faq` for more journal subscription information. Each number may be ordered separately; please specify number when ordering an individual number.

Back number information. For back issues see `www.ams.org/backvols`.

Subscriptions and orders should be addressed to the American Mathematical Society, P. O. Box 845904, Boston, MA 02284-5904 USA. *All orders must be accompanied by payment.* Other correspondence should be addressed to 201 Charles Street, Providence, RI 02904-2213 USA.

Memoirs of the American Mathematical Society (ISSN 0065-9266 (print); 1947-6221 (online)) is published bimonthly (each volume consisting usually of more than one number) by the American Mathematical Society at 201 Charles Street, Providence, RI 02904-2213 USA. Periodicals postage paid at Providence, RI. Postmaster: Send address changes to Memoirs, American Mathematical Society, 201 Charles Street, Providence, RI 02904-2213 USA.

This publication is indexed in *Mathematical Reviews®*, *Zentralblatt MATH*, *Science Citation Index®*, *Science Citation Index™-Expanded*, *ISI Alerting Services℠*, *SciSearch®*, *Research Alert®*, *CompuMath Citation Index®*, *Current Contents®/Physical, Chemical & Earth Sciences.*
This publication is archived in *Portico* and *CLOCKSS.*
Printed in the United States of America.

∞ The paper used in this book is acid-free and falls within the guidelines
established to ensure permanence and durability.
Visit the AMS home page at `https://www.ams.org/`

10 9 8 7 6 5 4 3 2 1 30 29 28 27 26 25

Dedicated to the memory of my parents

Contents

Abstract

In this paper, we consider an isospectral deformation of infinite periodic band matrices with period n, with lower bandwidth equal to k and upper bandwidth equal to k'. We show that this flow, which is introduced in van Moerbeke and Mumford, translates into a corresponding flow on matrix loops in $\mathfrak{gl}(n, \mathbb{C})$ which is a natural extension of the periodic Kostant-Toda flow. We show that the coadjoint orbits through the submanifold of such matrix loops can be identified with those of a finite-dimensional Lie group which appears in the form of a semi-direct product. We characterize the generic coadjoint orbits in this submanifold and compute their dimension. We also establish the Liouville integrability of the flow on the generic coadjoint orbits via the construction of action-angle variables. Finally, we study the correspondence between an open collection of loops corresponding to the infinite periodic band matrices and the collection of extended algebro-geometric data. The additional variables in the extended algebro-geometric data allow us to set up a bijection between the two sets of objects, thus extending the dictionary in van Moerbeke and Mumford.

Received by the editor November 25, 2019, and, in revised form, September 15, 2022, and January 6, 2023.

Article electronically published on June 10, 2025.

DOI: https://doi.org/10.1090/memo/1566

2020 *Mathematics Subject Classification.* Primary 37J35; Secondary 17B80, 14H70.

Key words and phrases. Liouville integrability, coadjoint orbits, algebraic curves with singularities, infinite periodic band matrices, periodic Toda flows.

The author is affiliated with the Department of Mathematics, Pennsylvania State University, University Park, Pennsylvania 16802.

The author would like to thank Camille Laurent-Gengoux for informing him of the work [**BA**]. He also gratefully acknowledges the support from the Simons Foundation through grants #278994 and #585813. Finally, the author would like to express his sincere gratitude to the referee for the thorough, and very helpful work he did in reviewing the paper.

CHAPTER 1

Introduction

In the mid 1970s, the study of the Korteweg-de Vries equation [**DMN, MvM**] and the periodic Toda lattice [**DMN, vM**] have led to the development of an inverse spectral theory for the Hill's operator and a parallel theory for periodic Jacobi matrices. This success in the second order case led van Moerbeke and Mumford to ask if there exists a similar theory for higher order periodic difference operators [**vMM**]. To give a quick summary of two of the main results in [**vMM**], let $\mathbf{C}$ denote an infinite periodic band matrix of period n, with lower bandwidth k and upper bandwidth k' (see Appendix A for the definition), subject to the conditions that

$$1 \leq k, k' \leq n-1, k+k' < n. \tag{1.1}$$

In [**vMM**], the first main result is a one-to-one correspondence between two sets of data, namely, the set of regular infinite periodic band matrices $\mathbf{C}$, modulo conjugation by periodic diagonal matrices of period n, and corresponding algebro-geometric data consisting of a spectral curve, and a "regular point" on the Jacobi variety of the curve. Secondly, the authors show that linear flows on the Jacobi variety correspond to isospectral deformations of $\mathbf{C}$, given by[1]

$$\dot{\mathbf{C}} = [\, P(\mathbf{C}, \Sigma, \Sigma^{-1})^{+}, \mathbf{C}\,] \quad \text{or} \quad \dot{\mathbf{C}} = [\, P(\mathbf{C}, \Sigma, \Sigma^{-1})^{[+]}, \mathbf{C}\,], \tag{1.2}$$

where $P(w, z, z^{-1})$ is a polynomial in w, z and z^{-1}, $(\Sigma u)_j = u_{j+n}$ and where X^{+}, $X^{[+]}$ for any difference operator X is defined by (compare [**vMM**])[2]

$$(X^{+})_{ij} = \begin{cases} x_{ij} & \text{for } i \leq j \\ 0 & \text{for } j < i, \end{cases} \qquad (X^{[+]})_{ij} = \begin{cases} x_{ij} & \text{for } i < j \\ \frac{1}{2}x_{ii} & \text{for } i = j \\ 0 & \text{for } j < i \end{cases}. \tag{1.3}$$

A preliminary study of the complete integrability of the second type of flows is also carried out in [**vMM**], namely, the authors exhibit a Poisson structure with respect to which the equations

$$\dot{\mathbf{C}} = [\,(\Sigma^{-k}\mathbf{C}^{l})^{[+]}, \mathbf{C}\,] \tag{1.4}$$

are Hamiltonian. Furthermore, the family of Hamiltonians $\{\mathrm{Tr}(\Sigma^{-k}\mathbf{C}^{l})\}$ are shown to Poisson commute with each other. In spite of this, a proof of the Liouville integrability of these flows is unfortunately lacking. Indeed, not even the dimension of the phases spaces is known from [**vMM**].

In this work, instead of considering the second type of flows above, we will focus on the first type of flows. Without loss of generality, let us just take the isospectral

[1] The spectrum here is the collection of all Floquet multipliers.

[2] Note that we are defining X^{+} to be the upper triangular part here.

deformation of $\mathbf{C}$, given by the equation

$$\dot{\mathbf{C}} = [\,\mathbf{C}^+, \mathbf{C}\,]. \tag{1.5}$$

One of the goals of this work is to establish Liouville integrability of this equation for generic $\mathbf{C}$. The other objective, which is intimately related to integrability and can be achieved via a detailed study of integrability, is to improve on the dictionary in [**vMM**] (which we mentioned above) by removing the "modulo conjugation by periodic diagonal matrices of period n" restriction. There are at least two reasons why we study this type of flows first: (a) the equation (1.5) is related to the periodic Kostant-Toda flow [**BA**], which is an extension of the Lax equation of the periodic Toda lattice (in Kostant's realization) to bigger phase spaces, (b) it is simpler in some aspects than the other type of flows, yet it sheds light on what to expect in general. And with this introduction, let us now turn to a description of the periodic Kostant-Toda flow and its generalization, which is what we call the *periodic Toda flow* in this article.

To begin with, let us recall that the Toda lattice was first introduced by M. Toda in [**T**] and different versions of the original model have been investigated by various authors, for a variety of reasons. The fact is that the various versions of the Toda lattice are related to a multitude of topics in mathematics and physics and connections to new areas are still being made. Examples include the description of Langmuir waves in plasma physics [**M, ZMR**], nonlinear optics [**Ar**], DNA modeling [**MSC**], Toda field theory and affine Toda field theory [**MOP, Wi, OT1, OT2, UT**], harmonic maps [**Gu**], orthogonal polynomials [**Ap**], numerical linear algebra [**Sy, Wat, DLT2**], and quantum cohomology [**GK**], to name just a few.

If we consider the Toda lattice with a finite number of particles on the line, the resulting system is called the finite nonperiodic Toda lattice. In this case, the Hamiltonian system can be described by two distinct Lax representations, associated with two different splittings of the same finite-dimensional Lie algebra into Lie subalgebras (see, for example, [**A, K2**]), and has given rise to natural extensions on bigger phase spaces that are still Liouville integrable [**DLNT, DLT1, EFS, GS**]. Indeed, one of the motivations behind making these extensions in [**DLT1**] has been the connection with certain eigenvalue algorithms in numerical linear algebra (see, for example, [**Sy, Wat, DLT2**]).

On the other hand, if we consider the Toda lattice with a finite number of particles on the circle, the corresponding system is called the periodic Toda lattice. In this case, a Lax pair for the equations of motion is due to Flaschka [**F**]. As in the finite nonperiodic case, the system also admits two different Lax representations, but now the splittings are connected with infinite dimensional loop algebras (see, for example, [**AvM**], [**BA**], and Appendix A). In [**BA**], Ben Abdeljelil introduced the periodic Kostant-Toda flow in the general context of complex simple Lie algebras $\mathfrak{g}$, by analogy with the Kostant-Toda flow in [**EFS**]. Moreover, in the so-called full case, she established the Liouville integrability of the system on the underlying Poisson manifold, in the sense of [**AvMV**]. Moreover, at the end of [**BA**], the author introduced a family of Poisson submanifolds parametrized by $1 \leq k \leq m_\ell$, where m_ℓ is the largest exponent of $\mathfrak{g}$, and made the conjecture that the system is also Liouville integrable on these submanifolds for $2 \leq k \leq m_\ell - 1$. (The $k = m_\ell$ case is the full case discussed in [**BA**] and the usual periodic Toda lattice corresponds to $k = 1$.) The solution of this conjecture for $\mathfrak{g} = \mathfrak{sl}(n, \mathbb{C})$ was announced in

[**LN1**] and full details was published in [**LN2**]. Since the author in [**BA**] does not explain the passage from the periodic Toda lattice to her Lax equation, nor does she mention the connection with flows on infinite periodic band matrices, for the reader's convenience, we discuss this in Appendix A. En route, we also take the opportunity to define what we mean by infinite periodic band matrices with bandwidths k and k'.

In this connection, it is perhaps worth to point out that the Kostant-Toda flow in [**EFS**] is just a restriction of the complex version of the LU flow in [**DLT1**] to the Poisson submanifold of lower Hessenberg matrices with 1's on the first superdiagonal. Likewise, the periodic Kostant-Toda flow in [**LN1, LN2**] is analogously the restriction of a more general system, which we call the *periodic Toda flow* here. In order to write down this system, let us introduce some notations. We will work with $\mathfrak{g} = \mathfrak{gl}(n, \mathbb{C})$ here, with the usual pairing $(\cdot, \cdot)$ given by $(X, Y) = \mathrm{tr}(XY)$, and let $\mathfrak{h}$ be its Cartan subalgebra consisting of diagonal matrices. For $1 \leq i, j \leq n$, let E_{ij} denote the $n \times n$ matrix whose (i, j) entry is 1 and zeros elsewhere. Then we have the following Lie subalgebras of $\mathfrak{g}$:

$$\mathfrak{n}_- = \oplus_{i>j}\mathbb{C}E_{ij}, \quad \mathfrak{n}_+ = \oplus_{i<j}\mathbb{C}E_{ij}, \quad \mathfrak{b}_\pm = \mathfrak{h} \oplus \mathfrak{n}_\pm. \tag{1.6}$$

Moreover, we have the (vector space) direct sum decomposition

$$\mathfrak{g} = \mathfrak{n}_- \oplus \mathfrak{b}_+. \tag{1.7}$$

If we let $\Pi_{\mathfrak{b}_+}$ and $\Pi_{\mathfrak{n}_-}$ be the projection operators to $\mathfrak{b}_+$ and $\mathfrak{n}_-$ relative to this splitting, then we have

$$\Pi_{\mathfrak{b}_+} A = A_+ + A_0, \quad \Pi_{\mathfrak{n}_-} A = A_-, \tag{1.8}$$

where A_+, A_0, and A_- are the strictly upper triangular, diagonal, and strictly lower triangular parts of A, respectively. If for each $j = 1, \ldots, n-1$, we put

$$\mathfrak{g}^{(j)} = \oplus_{i=1}^{n-j}\mathbb{C}E_{i,i+j}, \quad \mathfrak{g}^{(-j)} = \oplus_{i=1}^{n-j}\mathbb{C}E_{i+j,i}, \tag{1.9}$$

and let $\mathfrak{g}^{(0)} := \mathfrak{h}$, then the equation we are interested in is of the form

$$\dot{L}(z) = [L(z), B(z)], \tag{1.10}$$

with

$$L(z) = L_1 z + L_0 + L_{-1} z^{-1}, \quad B(z) = \Pi_{\mathfrak{n}_-} L_0 + L_{-1} z^{-1}, \tag{1.11}$$

where

$$L_0 \in \oplus_{j=-k}^{k'} \mathfrak{g}^{(j)}, L_{-1} \in \oplus_{j=1}^{k} \mathfrak{g}^{(n-j)}, L_1 \in \oplus_{j=1}^{k'} \mathfrak{g}^{(-(n-j))}, \tag{1.12}$$

and where the numbers k and k' are subjected to the conditions in (1.1). We will call the flow defined by (1.10), (1.11) and (1.12) the *periodic Toda flow on matrix loops of level* (k, k'). In the special case where $L_1 = E_{n1}$, $k' = 1$ with 1's on the first superdiagonal of L_0, (1.10) reduces to the periodic Kostant-Toda flow on matrix loops of level k in [**LN2**] and the case $k = 1$ corresponds to the complex Toda lattice. The relation between (1.5) and (1.10) will be established in Chapter 2.

We now describe the main results of the paper. In Chapter 2, we begin by describing the Floquet matrix of the infinite periodic band matrix **C**. Then we establish the relationship between (1.5) and (1.10). More precisely, we have the following result.

THEOREM A.

(a) *If* $\mathbf{C}$ *is an infinite periodic band matrix of period* n*, with lower bandwidth equal to* k *and upper bandwidth equal to* k'*, where* k *and* k' *are subjected to* (1.1)*, then its Floquet matrix* $L(z)$ *is the matrix loop given by* $L(z) = L_1 z + L_0 + L_{-1}z^{-1}$*, where* L_{-1}*,* L_0 *and* L_1 *satisfy the conditions in* (1.12)*. (And in the convention in Section* 2*, where row and column indices run from* 0 *to* $n-1$*, we have to shift the indices of summation by* -1 *in the definitions in* (1.9) *and we define* $\mathfrak{h} = \oplus_{i=0}^{n-1} E_{ii}$*.)*
(b) *If* $\mathbf{C}$ *evolves according to* (1.5)*, then its Floquet matrix* $L(z)$ *satisfies* (1.10) *with* $B(z)$ *given by* (1.11)*.*

We conclude this section with a result which will play an important role in studying a certain subset of the angle variables later on.

In Chapter 3, we introduce the Lie groups and and Lie algebras which are relevant in our investigation. We begin with the connected and simply-connected Lie groups G, $N_\pm$ and $B_\pm$ which integrate respectively the Lie algebras $\mathfrak{g}$, $\mathfrak{n}_\pm$ and $\mathfrak{b}_\pm$. Then we introduce the loop group $LG = C^\infty(S^1, G)$ and the loop algebra $L\mathfrak{g} = C^\infty(S^1, \mathfrak{g})$. We will denote by LG_+ (resp. LG_-) the Lie subgroup of LG consisting of loops g which extend analytically to the interior (resp. exterior) of the unit circle and we let $L\mathfrak{g}_+$ (resp. $L\mathfrak{g}_-$) be its corresponding Lie algebra. Then

$$\begin{aligned} \widetilde{N}_- &= \{g \in LG_- \mid g(\infty) \in N_-\}, \\ \widetilde{B}_+ &= \{g \in LG_+ \mid g(0) \in B_+\} \end{aligned} \tag{1.13}$$

are Lie subgroups of $LG_\pm$ which integrate

$$\begin{aligned} \widetilde{\mathfrak{n}_-} &= \{X \in L\mathfrak{g}_- \mid X(\infty) \in \mathfrak{n}_-\}, \\ \widetilde{\mathfrak{b}_+} &= \{X \in L\mathfrak{g}_+ \mid X(0) \in \mathfrak{b}_+\}, \end{aligned} \tag{1.14}$$

respectively and we have the splitting $L\mathfrak{g} = \widetilde{\mathfrak{n}}_- \oplus \widetilde{\mathfrak{b}}_+$. Put

$$\widetilde{G} = \{g \in LG \mid g = g_+ g_-^{-1}, \text{ where } g_+ \in \widetilde{B}_+, g_- \in \widetilde{N}_-\} \tag{1.15}$$

and endow it with a Lie group structure with the multiplication

$$g * h = g_+ h g_-^{-1}, \quad g, h \in \widetilde{G}. \tag{1.16}$$

Then the Lie algebra of $(\widetilde{G}, *)$ is $\widetilde{\mathfrak{g}} = \widetilde{\mathfrak{b}}_+ \ominus \widetilde{\mathfrak{n}}_-$. Note that $\widetilde{\mathfrak{g}} = (L\mathfrak{g}, [\cdot,\cdot]_R)$, where $[\cdot,\cdot]_R$ is the R-bracket with

$$R = \Pi_{\widetilde{\mathfrak{b}}_+} - \Pi_{\widetilde{\mathfrak{n}}_-}, \tag{1.17}$$

where $\Pi_{\widetilde{\mathfrak{b}}_+}$ and $\Pi_{\widetilde{\mathfrak{n}}_-}$ are the projection maps to $\widetilde{\mathfrak{b}}_+$ and $\widetilde{\mathfrak{n}}_-$ relative to the splitting above for $L\mathfrak{g}$. We will use the following nondegenerate ad-invariant pairing on $L\mathfrak{g}$:

$$(X, Y)_{L\mathfrak{g}} = \oint_{|z|=1} (X(z), Y(z)) \frac{dz}{2\pi i z}, \tag{1.18}$$

where $(\cdot,\cdot)$ is the usual pairing on $\mathfrak{g}$. We will also identify $\widetilde{\mathfrak{g}}^*$, the algebraic dual of $\widetilde{\mathfrak{g}}$, with $L\mathfrak{g}$ through this pairing. Now let

(1.19)

$$\begin{aligned} \mathcal{S} &= \{L(z) = L_1 z + L_0 + L_{-1} z^{-1} \mid L_0, L_1, L_{-1} \in \mathfrak{g}\}, \\ \mathcal{T}^{(k,k')} &= \{L \in \mathcal{S} \mid L_0 \in \oplus_{j=-k}^{k'} \mathfrak{g}^{(j)}, L_{-1} \in \oplus_{j=1}^{k} \mathfrak{g}^{(n-j)}, L_1 \in \oplus_{j=1}^{k'} \mathfrak{g}^{(-(n-j))}\}. \end{aligned}$$

Then both $\mathcal{S}$ and $\mathcal{T}^{(k,k')}$ are invariant under the coadjoint action of $\widetilde{G}$. Indeed, the coadjoint orbits of $\widetilde{G}$ through loops in $\mathcal{S}$ can be identified with those of a finite dimensional Lie group. To describe this, recall that we have the splitting in (1.7), therefore the operator

$$J = \Pi_{\mathfrak{b}_+} - \Pi_{\mathfrak{n}_-} \tag{1.20}$$

satisfies the modified Yang-Baxter equation (mYBE). Let $[\cdot,\cdot]_J$ denote the corresponding R-bracket and let $\mathfrak{g}_J = (\mathfrak{g}, [\cdot,\cdot]_J)$. Then

$$G_J = \{b_+ b_-^{-1} \mid b_+ \in B_+, b_- \in N_-\} \tag{1.21}$$

with multiplication defined by

$$b \cdot b' = b_+ b' b_-^{-1} \tag{1.22}$$

is a Lie group that integrates $\mathfrak{g}_J$. Now the group G_J acts on $\mathfrak{g} \oplus \mathfrak{g}$ via the formula

$$\rho(b)(\xi, \eta) = (Ad_{b_+}\xi, Ad_{b_-}\eta). \tag{1.23}$$

Hence we can define the semidirect product Lie group $S = G_J \ltimes_\rho (\mathfrak{g} \oplus \mathfrak{g})$ with multiplication given by

$$(b, (\xi, \eta))(b', (\xi', \eta')) = (b \cdot b', (\xi, \eta) + \rho(b)(\xi', \eta')). \tag{1.24}$$

If $\mathfrak{s}$ is the Lie algebra of S, we will identify its dual $\mathfrak{s}^*$ with $\mathfrak{g} \times (\mathfrak{g} \oplus \mathfrak{g})$ through the following pairing on $\mathfrak{g} \times (\mathfrak{g} \oplus \mathfrak{g})$:

$$\langle (u, (\xi, \eta)), (u', (\xi', \eta')) \rangle = (u, u') + ((\xi, \eta), (\xi', \eta')). \tag{1.25}$$

We have the following result.

THEOREM B1.
(a) *$\mathcal{S}$ is invariant under $Ad^*_{\widetilde{G}}(g^{-1})$ for any $g \in \widetilde{G}$. Moreover, if $L \in \mathcal{S}$, $L(z) = L_1 z + L_0 + L_{-1}z^{-1}$, then we can identify the coadjoint orbit $\mathcal{O}_L = \{Ad^*_{\widetilde{G}}(g^{-1})L \mid g \in \widetilde{G}\}$ with the coadjoint orbit of S through $(L_0, (L_{-1}, L_1)) \in \mathfrak{s}^*$.*
(b) *For each $1 \leq k, k' \leq n-1$, $\mathcal{T}^{(k,k')}$ is invariant under $Ad^*_{\widetilde{G}}(g^{-1})$ for each $g \in \widetilde{G}$. Equivalently, if*

$$(L_0, (L_{-1}, L_1)) \in \oplus_{j=-k}^{k'} \mathfrak{g}^{(j)} \times \left(\oplus_{j=1}^{k} \mathfrak{g}^{(n-j)} \oplus \oplus_{j=1}^{k'} \mathfrak{g}^{(-(n-j))}\right), \tag{1.26}$$

then so is $Ad^*_S(b, (\xi, \eta))^{-1}(L_0, (L_{-1}, L_1))$ for any $(b, (\xi, \eta)) \in S$.
(c) *The Lax equation (1.10) is Hamiltonian on the coadjoint orbits of $\widetilde{G}$ in $\mathcal{T}^{(k,k')}$ with Hamiltonian $H(L) = \frac{1}{2}(L, L)_{L\mathfrak{g}}$.*

The Poisson structures on the coadjoint orbits of $\widetilde{G}$ in $\mathcal{T}^{(k,k')}$ are the induced ones obtained by restricting the standard Lie-Poisson structure on the algebraic dual $\widetilde{\mathfrak{g}}^*$ of the Lie algebra $(\widetilde{\mathfrak{g}}, [\cdot,\cdot]_R))$, given by the expression (3.11) in Chapter 3. For simplicity, we will denote such induced Poisson structures again by $\{\cdot,\cdot\}$.

In the remainder of Chapter 3, our first task is to characterize the generic coadjoint orbits in $\mathcal{T}^{(k,k')}$, which we will identify with

$$\oplus_{j=-k}^{k'} \mathfrak{g}^{(j)} \times \left(\oplus_{j=1}^{k} \mathfrak{g}^{(n-j)} \oplus \oplus_{j=1}^{k'} \mathfrak{g}^{(-(n-j))}\right). \tag{1.27}$$

To do so, we introduce the following functions corresponding to the boundary elements of L_0, L_{-1} and L_1, for $(L_0,(L_{-1},L_1)) \in \mathcal{T}^{(k,k')}$:

$$
\begin{aligned}
&x_{k+j,j}(L_0,(L_{-1},L_1)) = (e_{k+j}, L_0 e_j),\ j=1,\dots,n-k,\\
&y_{j,n-k+j}(L_0,(L_{-1},L_1)) = (e_j, L_{-1} e_{n-k+j}),\ j=1,\dots,k,\\
&x_{j,k'+j}(L_0,(L_{-1},L_1)) = (e_j, L_0 e_{k'+j}),\ j=1,\dots,n-k',\\
&z_{n-k'+j,j}(L_0,(L_{-1},L_1)) = (e_{n-k'+j}, L_1 e_j),\ j=1,\dots,k',
\end{aligned} \tag{1.28}
$$

where $e_1, \dots, e_n$ are the vectors in the canonical basis of $\mathbb{C}^n$ and where $(x,y) = \sum_{i=1}^{n} x_i y_i$ for $x, y \in \mathbb{C}^n$. (We will use this pairing in $\mathbb{C}^n$ throughout this work.) Let

$$
\begin{aligned}
P(L) &= \left(\prod_{j=1}^{n-k} x_{k+j,j} \prod_{j=1}^{k} y_{j,n-k+j}\right)(L),\\
P'(L) &= \left(\prod_{j=1}^{n-k'} x_{j,k'+j} \prod_{j=1}^{k'} z_{n-k'+j,j}\right)(L),
\end{aligned} \tag{1.29}
$$

and introduce the set

$$
\mathcal{G}' = \{L \in \mathcal{T}^{(k,k')} \mid P(L), P'(L) \neq 0\}. \tag{1.30}
$$

Let

$$
\begin{aligned}
&d = \gcd(n,k), \quad k = k_1 d, \quad n = n_1 d,\\
&d' = \gcd(n,k'), \quad k' = k_1' d', \quad n = n_1' d',
\end{aligned} \tag{1.31}
$$

we also define the following functions on $\mathcal{T}^{(k,k')}$:

(1.32)

$$
\begin{aligned}
\sigma_i &= \prod_{j=1}^{n_1-k_1} x_{k+i+(j-1)d,i+(j-1)d} \prod_{j=1}^{k_1} y_{i+(j-1)d,n-k+i+(j-1)d},\ i=1,\dots,d,\\
\sigma_i' &= \prod_{j=1}^{n_1'-k_1'} x_{i+(j-1)d',k'+i+(j-1)d'} \prod_{j=1}^{k_1'} z_{n-k'+i+(j-1)d',i+(j-1)d'},\ i=1,\dots,d'.
\end{aligned}
$$

Note that these functions correspond to quantities introduced in [**vMM**, Section 1] where they are used in defining the notion of *regular periodic difference operators*. The meaning of these functions in our theory in given in the next result.

THEOREM B2.

(a) *The functions* $x_{j,k'+j}$, $j=1, \dots, n-k'$, $z_{n-k'+j,j}$, $j=1, \dots, k'$ *on* $\mathcal{T}^{(k,k')}$ *are invariant under the coadjoint action of* S.

(b) *The set*

$$
\mathcal{G}' = \{L \in \mathcal{T}^{(k,k')} \mid P(L), P'(L) \neq 0\} \tag{1.33}
$$

is an open, dense subset of $\mathcal{T}^{(k,k')}$ *which is invariant under the coadjoint action of* S.

(c) *The functions* σ_i, $i=1, \dots, d$, σ_i', $i=1, \dots, d'$ *on* $\mathcal{T}^{(k,k')}$ *are invariant under the coadjoint action of* S.

From this result, the set

$$
\mathcal{G} = \{L \in \mathcal{G}' \mid \sigma_i(L) \neq \sigma_j(L), i \neq j, \sigma_i'(L) \neq \sigma_j'(L), i \neq j\} \tag{1.34}
$$

is an open, dense subset of $\mathcal{T}^{(k,k')}$ invariant under the coadjoint action of S.

DEFINITION 1.1. A loop $L \in \mathcal{T}^{(k,k')}$ is generic iff $L \in \mathcal{G}$. An orbit $\mathcal{O}_L$ of the coadjoint action of $\widetilde{G}$ through $L \in \mathcal{T}^{(k,k')}$ is generic iff the loop L is generic.

Since the characteristic polynomial $\det(L(z)-wI)$ is invariant under the Hamiltonian flow and provides us with a collection of conserved quantities, our second task is to spell out the form of this characteristic polynomial and to single out the coadjoint orbit invariants amongst the coefficients of this polynomial. For this purpose, we will introduce the following two sets which are characterized by arithmetic conditions involving the bandwidths, the exponents $m_i = i$ (with $d_i = m_i + 1$) and the Coxeter number $h = n$ of $\mathfrak{g} = \mathfrak{gl}(n, \mathbb{C})$:

$$\begin{aligned} E &= \left\{ r \mid \lfloor \frac{kd_r}{h} \rfloor = 1 + \lfloor \frac{km_r - 1}{h} \rfloor \right\}, \\ E' &= \left\{ r \mid \lfloor \frac{k'd_r}{h} \rfloor = 1 + \lfloor \frac{k'm_r - 1}{h} \rfloor \right\} \setminus \{ r \mid m_r = n_1' s, s = 1, \dots, d' - 1 \}. \end{aligned} \tag{1.35}$$

THEOREM B3.
(a) *For* $L \in \mathcal{T}^{(k,k')}$, *we have*

$$\det(L(z) - wI) = (-1)^n w^n + \sum_{r=0}^{n-1} \left(\sum_{j=-\lfloor \frac{k'd_r}{h} \rfloor}^{\lfloor \frac{kd_r}{h} \rfloor} I_{r,j}(L) z^{-j} \right) w^{n-1-r}. \tag{1.36}$$

(b) $I_{r,\lfloor \frac{kd_r}{h} \rfloor}$ *is a Casimir on* $\mathcal{T}^{(k,k')}$ *if* $r \in E$.
(c) $I_{r,-\lfloor \frac{k'd_r}{h} \rfloor}$ *is a Casmir on* $\mathcal{T}^{(k,k')}$ *if* $r \in E'$.
(d) $\mathrm{Card}(E) = k + d - 1$, $\mathrm{Card}(E') = k'$.

It should be remarked that the proof of part (b) and (c) of the above theorem is rather lengthy and involves heavy use of (elementary) inequalities.

In Section 4, our main concern is to compute the dimension of the generic coadjoint orbits which we defined in the last section. Here we use the well known fact that the coadjoint orbit of a Lie group through a point is a homogeneous space which can be identified with the Lie group modulo the isotropy subgroup through the point [**K1**]. Consequently we can compute the dimension of the coadjoint orbit through a point by computing the dimension of the isotropy subalgebra through that point. Our result is the following.

THEOREM C. *The dimension of the generic coadjoint orbits of* S *in* $\mathcal{G} \subset \mathcal{T}^{(k,k')}$ *is given by* $(n-1)(k+k') + d' - d$.

In Chapter 5, we begin with a simple result which counts the number of nontrivial integrals $I_{r,j}$ which are in involution. By using Theorem B3, a calculation shows that this number is given by

$$\frac{1}{2}[(n-1)(k+k') + d' - d] \tag{1.37}$$

and involution is guaranteed by classical r-matrix theory [**STS**]. Then we proceed to study the spectral curve $C(L)$ and to construct the angle variables. The method we use to study the spectral curve and its normalization $\widetilde{C}(L)$ in $\mathbb{CP}^2$ is by means of the Newton polygon of the polynomial $I(L, z, w) = \det(zL(z) - wI)$ which turns out to be the triangular region $\Delta(I)$ with vertices at $(0,n)$, $(n-k,0)$ and $(n+k',0)$. For generic choice of L (see Definition 1.2 below), $C(L)$ is irreducible and has non-ordinary singular points only at $P_0=[1:0:0]$ and $P_\infty=[0:1:0]$. Therefore, by the Baker-Khovanskii theorem [**B**, **Kho**], the genus of the nonsingular model $\widetilde{C}(L)$ of $C(L)$ is given by

$$g(\widetilde{C}(L)) = \mathrm{Card}(\Delta(I)^o\cap\mathbb{N}^2) = \frac{1}{2}[(n-1)(k+k')-d-d'+2], \tag{1.38}$$

and a basis of holomorphic 1-forms of $\widetilde{C}(L)$ is provided by

$$\omega_{ab} = \frac{z^{a-1}w^{b-1}}{\frac{\partial I}{\partial w}}\,dz,\ (a,b)\in\Delta(I)^o\cap\mathbb{N}^2. \tag{1.39}$$

Moreover, if $\phi:\widetilde{C}(L)\longrightarrow C(L)$ is the normalization of $C(L)$, we have

$$\begin{aligned}\phi^{-1}(P_0) &= \{P_{0,1},\dots,P_{0,d}\},\\ \phi^{-1}(P_\infty) &= \{P_{\infty,1},\dots,P_{\infty,d'}\}.\end{aligned} \tag{1.40}$$

Following [**vMM**], let $\bar{f}=(f_1,\dots,f_{n-1},z)^T$ be the eigenvector of $zL(z)$ with last component equal to z so that $zL(z)\bar{f}=w\bar{f}$. An important result in [**vMM**] is the following: every $L\in\mathcal{G}$ leads to n regular divisors $D^{(i)}(L)$ of degree $g=g(\widetilde{C}(L))$, $i=1,\dots,n$, such that

$$(f_i) = D^{(i)}(L)-D(L)-\sum_{a=1}^{i}P_{\infty,a}+\sum_{a=0}^{i-1}P_{0,a},\ \ D(L)=D^{(n)}(L), \tag{1.41}$$

where $P_{0,i+d}=P_{0,i}$ and $P_{\infty,i+d'}=P_{\infty,i}$ for all $i\in\mathbb{Z}$. Then for $L\in\mathcal{G}$ satisfying additional genericity assumptions (see Definition 1.2 below), we define

$$\phi_{ab}(L)=\int_{D_0}^{D(L)}\omega_{ab},\quad (a,b)\in\Delta(I)^o\cap\mathbb{N}^2, \tag{1.42}$$

where we simply pick $D_0=gP_{0,1}$.

THEOREM D1. *For $(a,b)\in\Delta(I)^o\cap\mathbb{N}^2$, $-\lfloor\frac{k'd_r}{h}\rfloor\le j\le\lfloor\frac{kd_r}{h}\rfloor$, $r=1,\dots,n-1$, we have*

$$\{\phi_{ab},I_{r,j}\}(L)=\delta_{r,n-b}\delta_{j,n-a-b}. \tag{1.43}$$

In the case when $d'=1$, the number of nontrivial integrals in involution is identical to the genus of $\widetilde{C}(L)$. In this case, the above result suffices to establish integrability in the sense of Liouville on the generic coadjoint orbits in $\mathcal{T}^{(k,k')}$. However, when $d'>1$, it is immediate from the above result that for $(a,b)\in\Delta(I)^o\cap\mathbb{N}^2$, we have

$$\{\phi_{ab},I_{n_1'\mu,-k_1'\mu}\}(L)=0\quad\text{for all } \mu=1,\dots,d'-1. \tag{1.44}$$

Thus for $\mu=1,\dots,d'-1$, the Hamiltonian flow generated by $I_{n_1'\mu,-k_1'\mu}$ is not moving on the Jacobi variety. Therefore we need to construct variables which move linearly under these flows. Instead of dealing with $I_{n_1'\mu,-k_1'\mu}$, we find it more convenient to work with an equivalent set of variables, defined as follows:

$$J_i^\infty=\sum_{\mu=1}^{d'-1}I_{n_1'\mu,-k_1'\mu}(\sigma_i')^{d'-1-\mu},\ i=1,\dots,d'-1. \tag{1.45}$$

The quantities which move linearly under the Hamiltonian flows generated by the J_i^∞ are associated with the points $P_{\infty,1}, \ldots, P_{\infty,d'}$ in $\widetilde{C}(L)$ which project to the singular point $P_\infty \in C(L)$. To define these quantities, introduce the normalized eigenvector

$$\widehat{v}(P) = \frac{\bar{f}}{z}(P) \tag{1.46}$$

with last component equal to 1. For generic L, $\widehat{v}(P_{\infty,i})$ exists and indeed is an eigenvector of the lower triangular matrix $L(z(P))^{n_1'} z(P)^{-k_1'} \,|_{P=P_{\infty,i}}$ with corresponding eigenvalue given by $\sigma_i'(L)$. Set $A = (a_{ij})_{i,j=1}^n := L(z=1)$. We have the following Poisson bracket relations.

THEOREM D2. *For $1 \le i \le d'-1$, $1 \le j \le d'-1$, we have*

$$\begin{aligned}\{\log \widehat{v}_{n-d'+i}(P_{\infty,i}), J_j^\infty\}(L) =& \frac{\partial J_i^\infty}{\partial a_{n-d'+i,n-d'+i}}\delta_{ij}\\ =&(-1)^{n-1}\prod_{\substack{a=1\\a\neq i}}^{d'}(\sigma_i'(L)-\sigma_a'(L))\delta_{ij}.\end{aligned} \tag{1.47}$$

The proof of this result is lengthy and involves some preparatory work. In particular, we have to study the asymptotic behaviour of various cofactors of the matrix loop $z(P)L(z(P)) - w(P)I$ as P approaches $P_{\infty,i}$. On the other hand, we find it necessary to bring into our analysis the eigenvector $\widehat{v}'(P)$ of $(z(P)L(z(P)))^T$ normalized with first component equal to 1. Finally, the Desnanot-Jacobi formula [**Br**] is also playing a role here.

By combining the results obtained previously, a proof of the functional independence of the integrals on an open, dense subset of each generic coadjoint orbit is now possible and allows us to draw the following conclusion.

THEOREM D3. *The periodic Toda flow with Hamiltonian $H(L) = \frac{1}{2}(L,L)_{L\mathfrak{g}}$ is Liouville integrable on the generic coadjoint orbits in $\mathcal{T}^{(k,k')}$ for all (k,k') satisfying $1 \le k, k' \le n-1, k+k' < n$.*

In Chapter 6, our goal is to study the correspondence between an open subset of loops $\mathcal{T}_*^{(k,k')} \subset \mathcal{T}^{(k,k')}$ and a collection of data which we call *extended algebro-geometric data*. It is in this final section that we improve on the dictionary in [**vMM**]. And what makes this possible is our results in the previous sections. The extended algebro-geometric data associated to a loop $L \in \mathcal{T}_*^{(k,k')}$ consists not only of the curve $C(L)$ and the divisor $D(L)$, but also

(a) the vector $(\widehat{v}_{n-d'+1}(P_{\infty,1}), \ldots, \widehat{v}_{n-1}(P_{\infty,d'-1}))$ of the first nonzero components of the eigenvectors $\widehat{v}(P_{\infty,1}), \ldots, \widehat{v}(P_{\infty,d'-1})$,

(b) the coadjoint orbit invariants given by the entries $a_{1,k'+1}, \ldots, a_{n-k',n}$ on the k'-th superdiagonal of L_0, and the entries $a_{n-k'+1,1}, \ldots, a_{n-d',k'-d'}$ on the $(n-k')$-th subdiagonal of L_1.

More precisely, we define $\mathcal{T}_*^{(k,k')}$ as follows.

DEFINITION 1.2. The open subset $\mathcal{T}_*^{(k,k')} \subset \mathcal{T}^{(k,k')}$ consists of those $L \in \mathcal{T}^{(k,k')}$ satisfying conditions (a) to (f) in Definition 6.1, the list being an extension of those in Theorem 5.6 by condition (f).

In essence, these conditions are just a gathering of the various assumptions which appear in Chapter 5. Let us elaborate on conditions (c) and (e). In a nutshell, these are the conditions which guarantee that the spectral curve $C(L)$ has only singular points at P_0 and P_∞ for generic choice of L. More precisely, condition (c) comes from applying Kronecker's method. For we have $I(L,z,0) = z^{n-k}\overline{I}_1(L,z)$, $\frac{\partial I}{\partial z}(L,z,0) = z^{n-k-1}\overline{I}_2(L,z)$, and $\frac{\partial I}{\partial w}(L,z,0) = z^{n-k}\overline{I}_3(L,z)$. Therefore, if for indeterminates u_i, v_i, $1 \le i \le 3$, we introduce the resultant

$$\text{Res}\left(\sum_{i=1}^{3} u_i\bar{I}_i, \sum_{i=1}^{3} v_i\bar{I}_i\right) = \sum_{\alpha,\beta} u_1^{\alpha_1}u_2^{\alpha_2}u_3^{\alpha_3}v_1^{\beta_1}v_2^{\beta_2}v_3^{\beta_3}R_{\alpha,\beta}(L), \tag{1.48}$$

then the system $\overline{I}_1(L,z) = \overline{I}_2(L,z) = \overline{I}(L,z) = 0$ has no solution iff $R_{\alpha,\beta}(L) \neq 0$ for all α, β. As for condition (e), it follows from the definition of the A-discriminant (see [**GKZ**] and the proof of Proposition 5.5 (c)) that $\Delta_A(I(L,\cdot,\cdot)) \neq 0$ iff the curve $I(L,z,w) = 0$ has no singularities in $(\mathbb{C}^*)^2$, where $A = \Delta(I) \cap \mathbb{N}^2$. These two conditions, together with condition (d), also play a role in checking the nondegeneracy of $I(L,z,w)$ with respect to the Newton polygon $\Delta(I)$.

On the other hand, the properties of the extended algebro-geometric data leads us to the following definition.

DEFINITION 1.3. An admissible set of data consists of

(a) a curve C in $\mathbb{P}^2(1,1,2)$ with two singular points $P_0 = [1:0:0]$, $P_\infty = [0:1:0]$ whose affine part is defined by the irreducible polynomial

$$I(z,w) = (-1)^n w^n + \sum_{r=0}^{n-1}\left(\sum_{j=-\lfloor \frac{k'd_r}{h}\rfloor}^{\lfloor \frac{kd_r}{h}\rfloor} I_{r,j}z^{r+1-j}\right)w^{n-1-r} \tag{1.49}$$

and whose normalization $\phi : \widetilde{C} \longrightarrow C$ is such that

$$\begin{aligned} \phi^{-1}(P_0) &= \{P_{0,1},\ldots,P_{0,d}\}, \\ \phi^{-1}(P_\infty) &= \{P_{\infty,1},\ldots,P_{\infty,d'}\}, \end{aligned} \tag{1.50}$$

where

$$\begin{aligned} P_{0,i} &= \langle c_i t^{n_1} + \cdots, t^{n_1-k_1}\rangle, \quad c_i^{n_1-k_1} = \frac{1}{\sigma_i}, 1 \le i \le d, \\ P_{\infty,j} &= \langle c_j' t^{-n_1'} + \cdots, t^{-(n_1'+k_1')}\rangle, \quad (c_j')^{(n_1'+k_1')} = \frac{1}{\sigma_j'}, 1 \le j \le d'. \end{aligned} \tag{1.51}$$

Here $\langle f(t), g(t)\rangle$ for a pair $(f(t), g(t))$ of meromorphic functions on an open neighborhood of $0 \in \mathbb{C}$ is the equivalence class containing (f,g) where the equivalence relation is defined in (6.32) (see [**Kir**]).

(b) a divisor D of degree equal to the genus of $\widetilde{C}$ which is *general* (i.e. $\dim \mathscr{L}(D) = 1$, where $\mathscr{L}(\mathscr{U})$ for a divisor $\mathscr{U}$ denotes the set of meromorphic functions g on $\widetilde{C}$ satisfying $(g) \ge -\mathscr{U}$), is *regular* with respect to $\{P_{0,i}\}_{i\in\mathbb{Z}}$ and $\{P_{\infty,j}\}_{j\in\mathbb{Z}}$ ($P_{0,i+d} = P_{0,i}$ and $P_{\infty,i+d'} = P_{\infty,i}$ for all i) in the sense of van Moerbeke and

Mumford [vMM], i.e., for all j, we have[1]

$$\dim \mathscr{L}\left(D+\sum_{i=1}^{j} P_{\infty,i}-\sum_{i=0}^{j} P_{0,i}\right)=0, \tag{1.52}$$

and moreover, $\operatorname{supp} D \cap \{P_{\infty,1},\dots,P_{\infty,d'}\}=\emptyset$,
(c) $(v_{n-d'+1},\dots,v_{n-1}) \in (\mathbb{C}^*)^{d'-1}$,
(d) $(a_{1,k'+1},\dots,a_{n-k',n}) \in (\mathbb{C}^*)^{n-k'}$, $(a_{n-k'+1,1},\dots,a_{n-d',k'-d'}) \in (\mathbb{C}^*)^{k'-d'}$.

We will denote the collection of all admissible sets of data by $\mathcal{D}$.

THEOREM E. *The map*

$$\begin{aligned}\mathcal{T}_*^{(k,k')} \ni L \mapsto &\text{ the spectral curve } C(L) \text{ with normalization } \phi : \widetilde{C}(L) \longrightarrow C(L),\\ &\text{the divisor } D(L), (\widehat{v}_{n-d'+1}(P_{\infty,1}),\dots,\widehat{v}_{n-1}(P_{\infty,d'-1})),\\ &(a_{1,k'+1},\dots,a_{n-k',n}),(a_{n-k'+1,1},\dots,a_{n-d',k'-d'})\end{aligned} \tag{1.53}$$

is a bijection onto $\mathcal{D}$.

The proof of this result divides naturally into two parts, the forward problem and the inverse problem. In the forward problem, we show that the map is one-to-one. And in the inverse problem, we show that the map is onto.

Finally, in Chapter 7, the last section, we show how to solve the Cauchy problem for (1.10) via Riemann-Hilbert factorization problems, as suggested by the group-theoretic framework. The result is the following.

THEOREM F. *Let* $L^0 \in \mathcal{T}^{(k,k')}$ *be the initial data for the Cauchy problem associated with* (1.10) *and let* $[0,T)$ *be the maximal interval of existence of its solution* $L(z,t)$. *Then for* $t \in [0,T)$, *there exist unique holomorphic matrix-valued functions*

$$\begin{aligned} g_+(\cdot,t) &: \mathbb{CP}^1 \setminus \{\infty\} \longrightarrow GL(n,\mathbb{C}),\\ g_-(\cdot,t) &: \mathbb{CP}^1 \setminus \{0\} \longrightarrow GL(n,\mathbb{C}),\end{aligned} \tag{1.54}$$

which are smooth in $t \in [0,T)$, *solve the factorization problem*

$$e^{tL^0(z)} = g_+(z,t)g_-(z,t)^{-1}, \quad z \in \mathbb{CP}^1 \setminus \{0,\infty\} \tag{1.55}$$

and satisfy

$$g_+(\cdot,t)|_{S^1} \in \widetilde{B}_+, \quad g_-(\cdot,t)|_{S^1} \in \widetilde{N}_-. \tag{1.56}$$

Moreover, for initial data $L^0 \in \mathcal{T}_*^{(k,k')}$, $g_\pm(z,t)$ *can be constructed by means of Riemann theta functions associated with the Riemann surface of the curve* $\widetilde{C}(L^0)$ *and the solution of the Cauchy problem is given by*

$$L(z,t) = g_\pm(z,t)^{-1}L^0(z)g_\pm(z,t). \tag{1.57}$$

Thus the last part of the paper is devoted to the construction of $g_\pm(z,t)$ via Riemann theta functions and it is here that we have to solve Riemann-Hilbert problems (cf. [RS], [DL]). We will not try to summarize the formulas here but refer the reader to (7.25), (7.27), (7.29) and (7.30).

We hope to return to the other type of flows in subsequent work.

[1] See Remark 6.5 (a) for the interpretation of $\sum_{i=1}^{j} P_{\infty,i}$ (resp. $\sum_{i=0}^{j} P_{0,i}$) for $j<1$ (resp. $j<0$).

CHAPTER 2

Preliminaries

Let $\mathbf{C}$ be the infinite periodic band matrix of period n with lower bandwidth k and upper bandwidth k' subject to the condition in the introduction, and let Σ be the operator on $l^\infty(\mathbb{Z})$ defined by $(\Sigma u)_j = u_{j+n}$, then $\mathbf{C}\Sigma = \Sigma\mathbf{C}$. Therefore, if for $z \in S^1$, we define

$$X_{(z)} = \{u \in l^\infty(\mathbb{Z}) \mid \Sigma u = zu\}, \tag{2.1}$$

then the finite dimensional space $X_{(z)}$ is invariant under Σ. A basis of $X_{(z)}$ is given by the vectors

$$\delta_i = \sum_{j=-\infty}^{\infty} z^j e_{i+jn}, \quad i = 0, \ldots, n-1, \tag{2.2}$$

where e_j is the vector in $l^\infty(\mathbb{Z})$ with j-th component equal to 1 and zeros elsewhere. By definition, the Floquet matrix $L(z)$ is the matrix of $\mathbf{C}|_{X_{(z)}}$ with respect to the ordered basis $(\delta_0, \ldots, \delta_{n-1})$, i.e., $L(z)$ is defined by means of the relations

$$\mathbf{C}\delta_j = \sum_{j=0}^{n-1} (L(z))_{ij}\delta_i, \quad j = 0, \ldots, n-1. \tag{2.3}$$

Note that in the convention of this chapter, row and column indices of $n \times n$ matrices will be from 0 to $n-1$. We will revert to our usual practice in subsequent chapters.

PROPOSITION 2.1. *$L(z)$ as a Laurent polynomial in z is given by*

$$L(z) = z^{-1}L_{-1} + L_0 + zL_1, \tag{2.4}$$

where

$$(L_0)_{ij} = \begin{cases} c_{ij} & -k \le j-i \le k', \\ 0, & \textit{otherwise}, \end{cases}$$
$$(L_{-1})_{ij} = \begin{cases} c_{i,j-n}, & n-k \le j-i \le n-1, n-k \le j \le n-1, \\ 0, & \textit{otherwise}, \end{cases} \tag{2.5}$$
$$(L_1)_{ij} = \begin{cases} c_{i,j+n}, & n-k' \le i-j \le n-1, 0 \le j < k', \\ 0, & \textit{otherwise}. \end{cases}$$

PROOF. For $0 \le j \le n-1$, we have

(2.6)
$$
\begin{aligned}
\mathbf{C}\delta_j &= \sum_{i=-\infty}^{\infty} z^i \left(\sum_{a=-k'}^{k} c_{j+a+in,j+in}\, e_{j+a+in} \right) \\
&= \sum_{a=-k'}^{k} c_{j+a,j} \left(\sum_{i=-\infty}^{\infty} z^i e_{j+a+in} \right) \\
&= \begin{cases} \sum_{a=-k'}^{k} c_{j+a,j}\delta_{j+a}, & k'-1 < j < n-k, \\ \sum_{a=-k'}^{n-1-j} c_{j+a,j}\delta_{j+a} + z^{-1} \sum_{a=n-j}^{k} c_{j+a-n,j-n}\delta_{j+a-n}, & n-k \le j \le n-1, \\ \sum_{a=-j}^{k} c_{j+a,j}\delta_{j+a} + z\sum_{a=-k'}^{-j-1} c_{j+a+n,j+n}\delta_{j+a+n}, & 0 \le j \le k'-1. \end{cases}
\end{aligned}
$$

The assertion then follows by changing the indices of summation. □

We next introduce the infinite band matrix $\mathbf{C}^+$ by

$$
(\mathbf{C}^+)_{ij} = \begin{cases} c_{ij} & \text{for } i \le j \\ 0 & \text{for } j < i, \end{cases} . \tag{2.7}
$$

PROPOSITION 2.2. *If* $\mathbf{C}$ *evolves according to*

$$
\dot{\mathbf{C}} = [\,\mathbf{C}^+, \mathbf{C}\,], \tag{2.8}
$$

then

$$
\dot{L}(z) = [\,L(z), B(z)\,], \tag{2.9}
$$

where $B(z)$ *is the loop specified in* (1.11) *and* (1.12). *Note that in the convention of this section,* $\mathfrak{g} = \oplus_{i,j=0}^{n-1}\mathbb{C}E_{ij}$, $\mathfrak{g}^{(j)} = \oplus_{i=0}^{n-1-j}\mathbb{C}E_{i,i+j}$, $\mathfrak{g}^{(-j)} = \oplus_{i=0}^{n-1-j}\mathbb{C}E_{i+j,i}$, *for* $j = 1, \ldots, n-1$, *and* $\mathfrak{g}^{(0)} = \oplus_{i=0}^{n-1}\mathbb{C}E_{ii}$.

PROOF. Let $\mathbf{C}^- = \mathbf{C} - \mathbf{C}^+$, then the equation for $\mathbf{C}$ is equivalent to $\dot{\mathbf{C}} = [\,\mathbf{C}, \mathbf{C}^-\,]$. Let $Q(z)$ be the matrix of $\mathbf{C}^-|_{X_{(z)}}$ with respect to the ordered basis $(\delta_0, \ldots, \delta_{n-1})$. To establish the assertion, it suffices to show that $Q(z) = B(z)$. By a direct calculation, for $0 \le j \le n-1$, we have

$$
\begin{aligned}
\mathbf{C}^-\delta_j &= \sum_{i=-\infty}^{\infty} z^i \mathbf{C}^- e_{j+in} \\
&= \sum_{a=1}^{k} c_{j+a,j} \left(\sum_{i=-\infty}^{\infty} z^i e_{j+a+in} \right)
\end{aligned} \tag{2.10}
$$

We divide into two cases. For $j < n-k$, we have

$$
\begin{aligned}
\mathbf{C}^-\delta_j &= \sum_{a=1}^{k} c_{j+a,j}\delta_{j+a} \\
&= \sum_{i=j+1}^{j+k} c_{i,j}\delta_i.
\end{aligned} \tag{2.11}
$$

On the other hand, for $n-k \le j \le n-1$, we find

$$
\begin{aligned}
\mathbf{C}^{-}\delta_j & \\
&= \sum_{a=1}^{n-1-j} c_{j+a,j}\delta_{j+a} + z^{-1}\sum_{a=n-j}^{k} c_{j+a-n,j-n}\delta_{j+a-n} \\
&= \sum_{i=j+1}^{n-1} c_{ij}\delta_i + z^{-1}\sum_{i=0}^{j-(n-k)} c_{i,j-n}\delta_i.
\end{aligned} \tag{2.12}
$$

Hence the assertion follows from Proposition 2.1 and the definition of $B(z)$. □

Now let

$$
\begin{aligned}
d &= \gcd(n,k), \quad k = k_1 d, \quad n = n_1 d, \\
d' &= \gcd(n,k'), \quad k' = k_1' d', \quad n = n_1' d'.
\end{aligned} \tag{2.13}
$$

Our next result is a basic property of $\mathbf{C}$ which is important in the study of a certain subset of the angle variables later on.

PROPOSITION 2.3. *$\mathbf{C}^{n_1'}\Sigma^{-k_1'}$ is lower triangular. Consequently,*

$$
\mathbf{C}^{n_1'}\Sigma^{-k_1'}\delta_j = \sum_{a=0}^{n-j-1} \ell_{j+a,j}\delta_{j+a} + z^{-1}\sum_{a=n-j}^{2n-1-j} \ell_{j+a,j}\delta_{j+a-n} + O(z^{-2}) \tag{2.14}
$$

for $j = 0, \ldots, n-1$.

PROOF. Since $\mathbf{C}^{n_1'}$ has lower bandwidth equal to $n_1'k$ and upper bandwidth equal to $n_1'k'$, it is clear that $\mathbf{C}^{n_1'}\Sigma^{-k_1'}$ is lower triangular. Indeed, the lower bandwidth of $\mathbf{C}^{n_1'}\Sigma^{-k_1'}$ is given by $n(k_1+k_1')$. Consequently,

$$
\begin{aligned}
\mathbf{C}^{n_1'}\Sigma^{-k_1'}\delta_j &= \sum_{i=-\infty}^{\infty} z^i \sum_{a=j+in}^{\infty} \ell_{a,j+in} e_a \\
&= \sum_{i=-\infty}^{\infty} z^i \sum_{a=0}^{n(k_1+k_1')} \ell_{j+in+a,j+in} e_{j+in+a} \\
&= \sum_{a=0}^{n(k_1+k_1')} \ell_{j+a,j}\left(\sum_{i=-\infty}^{\infty} z^i e_{j+a+in}\right) \\
&= \sum_{a=0}^{n-j-1} \ell_{j+a,j}\delta_{j+a} + z^{-1}\sum_{a=n-j}^{2n-j-1} \ell_{j+a,j}\delta_{j+a-n} + O(z^{-2}).
\end{aligned} \tag{2.15}
$$

□

COROLLARY 2.4. *In the notation of the above proposition, we have*

$$
\lim_{z\to\infty} L(z)^{n_1'} z^{-k_1'} = \sum_{j=0}^{n-1}\sum_{a=0}^{n-j-1} \ell_{j+a,j} E_{j+a,j}. \tag{2.16}
$$

Thus $\lim_{z\to\infty} L(z)^{n_1'} z^{-k_1'}$ is lower triangular.

PROOF. From the definition of Σ, we have

$$\Sigma^{-k'_1}\delta_j = z^{-k'_1}\delta_j. \tag{2.17}$$

Hence it follows from the definition of the Floquet matrix $L(z)$ that

$$\mathbf{C}^{n'_1}\Sigma^{-k'_1}\delta_j = \sum_{i=0}^{n-1}(L(z)^{n'_1}z^{-k'_1})_{ij}\delta_i. \tag{2.18}$$

Equate (2.18) with (2.15) and let $z \to \infty$, we conclude that $\lim_{z\to\infty} L(z)^{n'_1}z^{-k'_1}$ exists and moreover is given by (2.16). □

REMARK 2.5. Proposition 2.1 gives the relationship between the infinite periodic band matrix $\mathbf{C}$ and its corresponding Floquet matrix $L(z)$. From (2.5), it is clear that $L_0 \in \oplus_{j=-k}^{k'}\mathfrak{g}^{(j)}$, $L_{-1} \in \oplus_{j=1}^{k}\mathfrak{g}^{(n-j)}$, and $L_1 \in \oplus_{j=1}^{k'}\mathfrak{g}^{(-(n-j))}$. Conversely, given a loop $L(z) = L_1 z + L_0 + L_{-1}z^{-1}$, with L_0, L_{-1} and L_1 as above, we can reverse the arguments to obtain the corresponding infinite periodic band matrix $\mathbf{C}$. We can also show that if $L(z)$ satisfies (2.9) where $B(z)$ is the loop specified in (1.11) and (1.12), then the corresponding $\mathbf{C}$ satisfies (2.8). We will leave the details to the interested reader.

CHAPTER 3

Lie Groups and Coadjoint Orbits

Let $G, N_{\pm}, B_{\pm}$ be the connected and simply-connected Lie groups which integrate respectively the Lie algebras $\mathfrak{g}, \mathfrak{n}_{\pm}$ and $\mathfrak{b}_{\pm}$. In addition to the finite dimensional Lie algebras and Lie groups, we will introduce the loop algebra $L\mathfrak{g} = C^{\infty}(S^1, \mathfrak{g})$ and the loop group $LG = C^{\infty}(S^1, G)$. We will denote by LG_+ (resp. LG_-) the Lie subgroup of LG consisting of loops g which extend analytically to the interior (resp. exterior) of the unit circle and we let $L\mathfrak{g}_+$ (resp. $L\mathfrak{g}_-$) be its corresponding Lie algebra. Then

$$\begin{aligned} \widetilde{N}_- &= \{g \in LG_- \mid g(\infty) \in N_-\}, \\ \widetilde{B}_+ &= \{g \in LG_+ \mid g(0) \in B_+\} \end{aligned} \tag{3.1}$$

are Lie subgroups of LG_- and LG_+ which integrate the Lie subalgebras

$$\begin{aligned} \widetilde{\mathfrak{n}}_- &= \{X \in L\mathfrak{g}_- \mid X(\infty) \in \mathfrak{n}_-\}, \\ \widetilde{\mathfrak{b}}_+ &= \{X \in L\mathfrak{g}_+ \mid X(0) \in \mathfrak{b}_+\} \end{aligned} \tag{3.2}$$

respectively. The loop algebra $L\mathfrak{g}$ has the following (vector space) direct sum decomposition

$$L\mathfrak{g} = \widetilde{\mathfrak{n}}_- \oplus \widetilde{\mathfrak{b}}_+. \tag{3.3}$$

If for $X \in \mathfrak{g}$, we write $X(z) = X_-(z) + X_0 + X_+(z)$, where $X_- \in L\mathfrak{g}_-$ with $X_-(\infty) = 0$, $X_+ \in L\mathfrak{g}_+$ with $X(0) = 0$, and $X_0 \in \mathfrak{g}$, then the projection maps relative to the splitting in (3.3) are given by

$$\begin{aligned} (\Pi_{\widetilde{\mathfrak{n}}_-} X)(z) &= X_-(z) + \Pi_{\mathfrak{n}_-} X_0, \\ (\Pi_{\widetilde{\mathfrak{b}}_+} X)(z) &= X_+(z) + \Pi_{\mathfrak{b}_+} X_0, \end{aligned} \tag{3.4}$$

where $\Pi_{\mathfrak{n}_-}$ and $\Pi_{\mathfrak{b}_+}$ are projection maps to $\mathfrak{n}_-$ and $\mathfrak{b}_+$ relative to the splitting $\mathfrak{g} = \mathfrak{n}_- \oplus \mathfrak{b}_+$ (see (1.8)).

Now let

$$\widetilde{G} = \{\, g \in LG \mid g = g_+ g_-^{-1}, \text{ where } g_+ \in \widetilde{B}_+, g_- \in \widetilde{N}_-\}. \tag{3.5}$$

Then following the procedure in [**DLT1**], we can endow $\widetilde{G}$ with a Lie group structure with the multiplication

$$g * h = g_+ h g_-^{-1}, g, h \in \widetilde{G} \tag{3.6}$$

and we can show that $(\widetilde{G}, *)$ is a Lie group which integrates $\widetilde{\mathfrak{g}} := \widetilde{\mathfrak{b}}_+ \ominus \widetilde{\mathfrak{n}}_-$, the Lie algebra anti-direct sum of $\widetilde{\mathfrak{b}}_+$ and $\widetilde{\mathfrak{n}}_-$. Note that we can also regard $\widetilde{\mathfrak{g}}$ as $(L\mathfrak{g}, [\cdot,\cdot]_R)$ where $[\cdot,\cdot]_R$ is the R-bracket with R defined by

$$R = \Pi_{\widetilde{\mathfrak{b}}_+} - \Pi_{\widetilde{\mathfrak{n}}_-}. \tag{3.7}$$

We will use the following nondegenerate ad-invariant pairing on $L\mathfrak{g}$:

$$(X,Y)_{L\mathfrak{g}} = \oint_{|z|=1} (X(z),Y(z))\,\frac{dz}{2\pi i z}, \tag{3.8}$$

where $(\cdot,\cdot)$ is the usual pairing on $\mathfrak{g}$.

Now let $\widetilde{\mathfrak{g}}^*$ be the algebraic dual of $\widetilde{\mathfrak{g}}$ which we will identify with $L\mathfrak{g}$ through the pairing in (3.8). Then the coadjoint action of $\widetilde{G}$ on $\widetilde{\mathfrak{g}}^*$ is given by

$$Ad^*_{\widetilde{G}}(g^{-1})X = \Pi_{\widetilde{\mathfrak{b}}_+^\perp} Ad_{g_-} X + \Pi_{\widetilde{\mathfrak{n}}_-^\perp} Ad_{g_+} X, \tag{3.9}$$

where

$$\begin{aligned} \widetilde{\mathfrak{b}}_+^\perp &= \{\, X = X_+ + X_0 \in L\mathfrak{g} \mid X_0 \in \mathfrak{b}_+^\perp = \mathfrak{n}_+ \,\},\\ \widetilde{\mathfrak{n}}_-^\perp &= \{\, X = X_- + X_0 \in L\mathfrak{g} \mid X_0 \in \mathfrak{n}_-^\perp = \mathfrak{b}_- \,\}, \end{aligned} \tag{3.10}$$

and $\Pi_{\widetilde{\mathfrak{b}}_+^\perp}, \Pi_{\widetilde{\mathfrak{n}}_-^\perp}$ are the projection maps relative to the splitting $L\mathfrak{g} = \widetilde{\mathfrak{b}}_+^\perp \oplus \widetilde{\mathfrak{n}}_-^\perp$.

Finally the Lie-Poisson structure on $\widetilde{\mathfrak{g}}^*$ is given by the usual formula:

$$\{F_1,F_2\}(X) = (X, [dF_1(X), dF_2(X)]_R)_{L\mathfrak{g}}. \tag{3.11}$$

With this set up, it is now easy to check that if we take

$$H(X) = \frac{1}{2}(X,X)_{L\mathfrak{g}}, \tag{3.12}$$

then the Hamiltonian equation generated by H with respect to the Poisson structure $\{\cdot,\cdot\}$ is given by

$$\begin{aligned} \dot{X}(z) &= [X(z), (\Pi_{\widetilde{\mathfrak{n}}_-} X)(z)]\\ &= [X(z), X_-(z) + \Pi_{\mathfrak{n}_-} X_0] \end{aligned} \tag{3.13}$$

and if we take L as defined in (1.11), (1.12), the corresponding equation (1.10) is generated by $H(L) = \frac{1}{2}(L,L)_{L\mathfrak{g}}$ in the induced Poisson structure. (We will check below that (1.12) indeed defines a submanifold $\mathcal{T}^{(k,k')}$ invariant under the coadjoint action of $\widetilde{G}$.)

We next introduce the submanifold

$$\mathcal{S} = \{L_1 z + L_0 + L_{-1} z^{-1} \mid L_0, L_1, L_{-1} \in \mathfrak{g}\} \tag{3.14}$$

of $\widetilde{\mathfrak{g}}^*$ which clearly contains the phase spaces of (1.10) for $1 \le k, k' \le n-1$, $k + k' < n$. Our immediate goal is to show that the $\mathcal{S}$ is invariant under the coadjoint action of $\widetilde{G}$.

PROPOSITION 3.1. *$\mathcal{S}$ is invariant under $Ad^*_{\widetilde{G}}(g^{-1})$ for any $g \in \widetilde{G}$. Indeed, if $L \in \mathcal{S}$, then*

$$\begin{aligned} (Ad^*_{\widetilde{G}}(g^{-1})L)(z) &= (Ad_{g_-(\infty)} L_1)\, z + \Pi_{\mathfrak{n}_+}(Ad_{g_-(\infty)} L_0 + [\, Ad_{g_-(\infty)} L_1, b_{-1}\,])\\ &\quad + \Pi_{\mathfrak{b}_-}(Ad_{g_+(0)} L_0 + [\, c_1, Ad_{g_+(0)} L_{-1}\,]) + (Ad_{g_+(0)} L_{-1})\, z^{-1}, \end{aligned} \tag{3.15}$$

where $\Pi_{\mathfrak{n}_+}$ and $\Pi_{\mathfrak{b}_-}$ are projection operators to $\mathfrak{n}_+$ and $\mathfrak{b}_-$ respectively relative to the splitting

$$\mathfrak{g} = \mathfrak{n}_+ \oplus \mathfrak{b}_-, \tag{3.16}$$

and

$$(3.17)\qquad \begin{aligned} b_{-1} &= \lim_{z\to\infty} z^2 T_{g_-(z)} r_{g_-(z)^{-1}} g_-'(z), \\ c_1 &= \lim_{z\to 0} T_{g_+(z)} r_{g_+(z)^{-1}} g_+'(z). \end{aligned}$$

PROOF. Let $L(z) = L_1 z + L_0 + L_{-1}z^{-1}$, and let $g = g_+ g_-^{-1} \in \widetilde{G}$, where $g_+ \in \widetilde{B}_+$, $g_- \in \widetilde{N}_-$. Write

$$(3.18)\qquad (Ad_{g_-} L)(z) = \sum_j X_j z^j.$$

From the form of L, we have

$$(3.19)\qquad \lim_{z\to\infty} Ad_{g_-(z)}(z^{-1}L(z)) = Ad_{g_-(\infty)} L_1.$$

Hence it follows that in (3.18) above, we must have $X_j = 0$ for $j > 1$ and moreover $X_1 = Ad_{g_-(\infty)} L_1$. Consequently,

$$(3.20)\qquad \left(\Pi_{\tilde{\mathfrak{b}}_+^{\perp}} Ad_{g_-} L\right)(z) = \Pi_{\mathfrak{n}_+} X_0 + \left(Ad_{g_-(\infty)} L_1\right) z.$$

where $\Pi_{\mathfrak{n}_+}$ (resp. $\Pi_{\mathfrak{b}_-}$) is the projection map to $\mathfrak{n}_+$ (resp. $\mathfrak{b}_-$) relative to the splitting $\mathfrak{g} = \mathfrak{n}_+ \oplus \mathfrak{b}_-$. In a similar way, write

$$(3.21)\qquad (Ad_{g_+} L)(z) = \sum_j Y_j z^j.$$

Then from

$$(3.22)\qquad \lim_{z\to 0} Ad_{g_+(z)}(zL(z)) = Ad_{g_+(0)} L_{-1},$$

we conclude that $Y_j = 0$ for $j < -1$ and $Y_{-1} = Ad_{g_+(0)} L_{-1}$. So

$$(3.23)\qquad \left(\Pi_{\tilde{\mathfrak{n}}_-^{\perp}} Ad_{g_+} L\right)(z) = \Pi_{\mathfrak{b}_-} Y_0 + (Ad_{g_+(0)} L_{-1}) z^{-1}.$$

Assembling the calculation, we find

$$(3.24)\qquad \left(Ad^*_{\widetilde{G}}(g^{-1})L\right)(z) = \left(Ad_{g_-(\infty)} L_1\right) z + \Pi_{\mathfrak{n}_+} X_0 + \Pi_{\mathfrak{b}_-} Y_0 + (Ad_{g_+(0)} L_{-1}) z^{-1}$$

and it remains to compute X_0 and Y_0 explicitly. First, it follows from our above analysis that

$$(3.25)\qquad \frac{d}{dz}(Ad_{g_-(z)}(z^{-1}L(z))) = \sum_{j=-\infty}^{0} (j-1) X_j z^{j-2}.$$

On the other hand, by differentiation, the left hand side of the above expression is equal to

$$(3.26)\qquad \begin{aligned} &\frac{d}{dz}(Ad_{g_-(z)}(z^{-1}L(z))) \\ &= Ad_{g_-(z)} \frac{d}{dz}(z^{-1}L(z)) + [\, T_{g_-(z)} r_{g_-(z)^{-1}} g_-'(z), Ad_{g_-(z)}(z^{-1}L(z)) \,]. \end{aligned}$$

If we multiply the above expression by z^2 and let $z \to \infty$, we find

$$(3.27)\qquad X_0 = Ad_{g_-(\infty)} L_0 + [\, Ad_{g_-(\infty)} L_1, b_{-1} \,],$$

where b_{-1} is given by the first line of (3.17). On the other hand, we have

$$\frac{d}{dz} Ad_{g_+(z)}(zL(z)) = \sum_{j=0}^{\infty}(j+1)Y_j z^j, \tag{3.28}$$

and by differentiation, the left hand side of the above expression is given by

$$\begin{aligned} &\frac{d}{dz} Ad_{g_+(z)}(zL(z)) \\ &= Ad_{g_+(z)}\frac{d}{dz}(zL(z)) + [\,T_{g_+(z)}r_{g_+(z)^{-1}}g_+'(z), Ad_{g_+(z)}(zL(z))\,]. \end{aligned} \tag{3.29}$$

Therefore, on taking the limit of the above expression as $z \to 0$, it follows from (3.28) and (3.29) that

$$Y_0 = Ad_{g_+(0)}L_0 + [\,c_1, Ad_{g_+(0)}L_{-1}\,], \tag{3.30}$$

where c_1 is given by the second line of (3.17). On substituting (3.27) and (3.30) into (3.24), we obtain (3.15), as asserted. □

We next show that the coadjoint orbits of $\widetilde{G}$ through the loops in $\mathcal{S}$ can be identified with those of a finite dimensional Lie group. For this purpose, recall again that $\mathfrak{g} = \mathfrak{n}_- \oplus \mathfrak{b}_+$. Therefore, the operator

$$J = \Pi_{\mathfrak{b}_+} - \Pi_{\mathfrak{n}_-} \tag{3.31}$$

satisfies the modified Yang-Baxter equation (mYBE). Let $[\cdot,\cdot]_J$ denote the corresponding R-bracket and let $\mathfrak{g}_J = (\mathfrak{g}, [\cdot,\cdot]_J)$. Then

$$G_J = \{b_+ b_-^{-1} \mid b_+ \in B_+, b_- \in N_-\} \tag{3.32}$$

with multiplication defined by

$$b \cdot b' = b_+ b' b_-^{-1} \tag{3.33}$$

is a Lie group that integrates $\mathfrak{g}_J$. Now the group G_J acts on $\mathfrak{g} \oplus \mathfrak{g}$ via the formula

$$\rho(b)(\xi, \eta) = (Ad_{b_+}\xi, Ad_{b_-}\eta). \tag{3.34}$$

Hence we can define the semidirect product Lie group $S = G_J \ltimes_\rho (\mathfrak{g} \oplus \mathfrak{g})$ with multiplication given by

$$(b, (\xi, \eta))(b', (\xi', \eta')) = (b \cdot b', (\xi, \eta) + \rho(b)(\xi', \eta')). \tag{3.35}$$

If $\mathfrak{s}$ is the Lie algebra of S and we identify its dual $\mathfrak{s}^*$ with $\mathfrak{g} \times (\mathfrak{g} \oplus \mathfrak{g})$ through the following pairing on $\mathfrak{g} \times (\mathfrak{g} \oplus \mathfrak{g})$:

$$\langle (u, (\xi, \eta)), (u', (\xi', \eta'))\rangle = (u, u') + ((\xi, \eta), (\xi', \eta')), \tag{3.36}$$

we have the following result.

PROPOSITION 3.2. *Let $L \in \mathcal{S}$, $L(z) = L_1 z + L_0 + L_{-1}z^{-1}$. Then we can identify the coadjoint orbit $\mathcal{O}_L = \{Ad^*_{\widetilde{G}}(g^{-1})L \mid g \in \widetilde{G}\}$ with the coadjoint orbit of S through $(L_0, (L_{-1}, L_1)) \in \mathfrak{s}^*$.*

PROOF. First, note that for the Lie group G_J, the exponential map and the adjoint action are given respectively by

$$\exp_{G_J}(tu) = e^{t\Pi_{\mathfrak{b}_+}u} e^{t\Pi_{\mathfrak{n}_-}u}, \quad u \in \mathfrak{g}_J, \tag{3.37}$$

$$Ad_{G_J}(b)\xi = Ad_{b_+}\Pi_{\mathfrak{b}_+}\xi + Ad_{b_-}\Pi_{\mathfrak{n}_-}\xi, \quad b \in G_J,\ \xi \in \mathfrak{g}_J. \tag{3.38}$$

Therefore,

$$\rho'(u)(\xi,\eta) = \frac{d}{dt}\Big|_{t=0}\rho(\exp_{G_J}(tu))(\xi,\eta) = ([\Pi_{\mathfrak{b}_+}u,\xi], -[\Pi_{\mathfrak{n}_-}u,\eta]). \tag{3.39}$$

Hence the adjoint action of S on $\mathfrak{s}$ is given by

$$\begin{aligned} &Ad_S(b,(\xi,\eta))(u,(\xi',\eta')) \\ &= (Ad_{G_J}(b)u, \rho(b)(\xi',\eta') - \rho'(Ad_{G_J}(b)u)(\xi,\eta)) \\ &= \left(Ad_{G_J}(b)u, (Ad_{b_+}\xi' - [Ad_{b_+}\Pi_{\mathfrak{b}_+}u,\xi], Ad_{b_-}\eta' + [Ad_{b_-}\Pi_{\mathfrak{n}_-}u,\eta])\right). \end{aligned} \tag{3.40}$$

On using this expression, a direct calculation using the pairing $\langle\cdot,\cdot\rangle$ on $\mathfrak{g}\times(\mathfrak{g}\oplus\mathfrak{g})$ above then yields

$$\begin{aligned} &Ad_S^*(b,(\xi,\eta))^{-1}(L_0,(L_{-1},L_1)) \\ &= (\,\Pi_{\mathfrak{n}_+}(Ad_{b_-}L_0 + [\eta, Ad_{b_-}L_1]) + \Pi_{\mathfrak{b}_-}(Ad_{b_+}L_0 + [\xi, Ad_{b_+}L_{-1}]), \\ &\quad (Ad_{b_+}L_{-1}, Ad_{b_-}L_1)\,). \end{aligned} \tag{3.41}$$

When we compare this expression with the corresponding one in Proposition 3.1, the assertion follows. □

We now introduce

$$\mathcal{T}^{(k,k')} = \{L\in\mathcal{S} \mid L_1 \in \oplus_{j=1}^{k'}\,\mathfrak{g}^{(-(n-j))}, L_0\in\oplus_{j=-k}^{k'}\,\mathfrak{g}^{(j)},\ L_{-1}\in\oplus_{j=1}^{k}\,\mathfrak{g}^{(n-j)}\}, \tag{3.42}$$

which is where $L(z)$ resides in equation (1.10). We will show that $\mathcal{T}^{(k,k')}$ is invariant under the coadjoint action of $\widetilde{G}$.

PROPOSITION 3.3. *For each $1\le k,k'\le n-1$, $\mathcal{T}^{(k,k')}$ is invariant under $Ad^*_{\widetilde{G}}(g^{-1})$ for each $g\in\widetilde{G}$. Equivalently, if*

$$(L_0,(L_{-1},L_1)) \in \oplus_{j=-k}^{k'}\mathfrak{g}^{(j)} \times \left(\oplus_{j=1}^{k}\mathfrak{g}^{(n-j)} \oplus \oplus_{j=1}^{k'}\mathfrak{g}^{(-(n-j))}\right), \tag{3.43}$$

then so is $Ad_S^*(b,(\xi,\eta))^{-1}(L_0,(L_{-1},L_1))$ for any $(b,(\xi,\eta))\in S$.

PROOF. Let $L\in\mathcal{T}^{(k,k')}$, $L(z) = L_1\,z + L_0 + L_{-1}\,z^{-1}$. Since $\mathcal{T}^{(k,k')}\subset\mathcal{S}$, it follows that $Ad_{\widetilde{G}}(g^{-1})L$ is given by the formula in (3.15). Therefore, it suffices to show that

$$\begin{aligned} &\Pi_{\mathfrak{b}_-}(Ad_{g_+(0)}L_0 + [\,c_1, Ad_{g_+(0)}L_{-1}\,]) + \Pi_{\mathfrak{n}_+}(Ad_{g_-(\infty)}L_0 + [\,Ad_{g_-(\infty)}L_1, b_{-1}\,]) \\ &\qquad \in \oplus_{j=-k}^{k'}\,\mathfrak{g}^{(j)}, \\ &Ad_{g_+(0)}L_{-1} \in \oplus_{j=1}^{k}\,\mathfrak{g}^{(n-j)}, Ad_{g_-(\infty)}L_1 \in \oplus_{j=1}^{k'}\,\mathfrak{g}^{(-(n-j))}, \end{aligned} \tag{3.44}$$

where $g_+(0)\in B_+$, $g_-(\infty)\in N_-$, Of these, the second line is obvious. From this, it follows that $[c_1, Ad_{g_+(0)}L_{-1}]\in\oplus_{j=-(k-1)}^{n-1}\mathfrak{g}^{(j)}$, $[b_{-1}, Ad_{g_-(\infty)}L_1]\in\oplus_{j=-(n-1)}^{k'-1}\mathfrak{g}^{(j)}$. On the other hand, from $g_+(0)\in B_+$, we have $Ad_{g_+(0)}L_0\in\oplus_{j=-k}^{n-1}\mathfrak{g}^{(j)}$, Therefore,

$$\Pi_{\mathfrak{b}_-}(Ad_{g_+(0)}L_0 + [\,c_1, Ad_{g_+(0)}L_{-1}\,]) \in \oplus_{j=-k}^{0}\mathfrak{g}^{(j)}. \tag{3.45}$$

Similarly, from $g_-(\infty)\in N_-$, it follows that $Ad_{g_-(\infty)}L_0\in\oplus_{j=0}^{k'}\mathfrak{g}^{(j)}$ and so

$$\Pi_{\mathfrak{n}_+}(Ad_{g_-(\infty)}L_0 + [\,Ad_{g_-(\infty)}L_1, b_1\,]) \in \oplus_{j=1}^{k'}\mathfrak{g}^{(j)}. \tag{3.46}$$

On combining (3.45) and (3.46), the first line in (3.44) follows. This completes the proof. □

We next characterize what we mean by the generic coadjoint orbits in $\mathcal{T}^{(k,k')}$. In view of our previous two propositions, we will state our next few results using the coadjoint action of the semidirect product S. Also, we will identify $\mathcal{T}^{(k,k')}$ with $\oplus_{j=-k}^{k'}\mathfrak{g}^{(j)} \times \left(\oplus_{j=1}^{k}\mathfrak{g}^{(n-j)} \oplus \oplus_{j=1}^{k'}\mathfrak{g}^{(-(n-j))}\right)$.

In order to define the generic elements in $\mathcal{T}^{(k,k')}$, we introduce the following functions corresponding to the boundary elements of L_0, L_{-1} and L_1, for $(L_0,(L_{-1},L_1)) \in \mathcal{T}^{(k,k')}$:

$$\begin{aligned}
&x_{k+j,j}(L_0,(L_{-1},L_1)) = (e_{k+j}, L_0 e_j),\ j=1,\dots,n-k,\\
&y_{j,n-k+j}(L_0,(L_{-1},L_1)) = (e_j, L_{-1} e_{n-k+j}),\ j=1,\dots,k,\\
&x_{j,k'+j}(L_0,(L_{-1},L_1)) = (e_j, L_0 e_{k'+j}),\ j=1,\dots,n-k',\\
&z_{n-k'+j,j}(L_0,(L_{-1},L_1)) = (e_{n-k'+j}, L_1 e_j),\ j=1,\dots,k'.
\end{aligned} \tag{3.47}$$

PROPOSITION 3.4. *For all* $L=(L_0,(L_{-1},L_1)) \in \mathcal{T}^{(k,k')}$, *and for all* $(b,(\xi,\eta)) \in S$, *we have*

(a) $x_{k+j,j}(Ad_S^*(b,(\xi,\eta))^{-1}(L)) = \frac{(e_{k+j},b_+e_{k+j})}{(e_j,b_+e_j)}\, x_{k+j,j}(L)$,
(b) $y_{j,n-k+j}(Ad_S^*(b,(\xi,\eta))^{-1}(L)) = \frac{(e_j,b_+e_j)}{(e_{n-k+j},b_+e_{n-k+j})}\, y_{j,n-k+j}(L)$,
(c) $x_{j,k'+j}(Ad_S^*(b,(\xi,\eta))^{-1}(L)) = x_{j,k'+j}(L)$,
(d) $z_{n-k'+j,j}(Ad_S^*(b,(\xi,\eta))^{-1}(L)) = z_{n-k'+j,j}(L)$.

PROOF.
(a) Since $(e_{k+j}, \Pi_{\mathfrak{n}_+}(Ad_{b_-}L_0 + [Ad_{b_-}L_0,\eta])e_j) = 0$, we have

$$x_{k+j,j}(Ad_S^*(b,(\xi,\eta))^{-1}(L)) = (e_{k+j}, (\Pi_{\mathfrak{b}_-}(Ad_{b_+}L_0 + [\xi, Ad_{b_+}L_{-1}]))e_j). \tag{3.48}$$

The remainder of the proof therefore proceeds exactly as in [**LN2**, Proposition 2.3 (a)].
(b) The argument can be reduced to the case in [**LN2**, Proposition 2.3 (b)].
(c) We have

$$\begin{aligned}
&x_{j,k'+j}(Ad_S^*(b,(\xi,\eta))^{-1}(L_0,(L_{-1},L_1))\\
=&(e_j, (\Pi_{\mathfrak{b}_+}(Ad_{b_-}L_0 + [\eta, Ad_{b_-}L_1]))e_{k'+j})\\
=&(e_j, (Ad_{b_-}L_0 + [\eta, Ad_{b_-}L_1])e_{k'+j}).
\end{aligned} \tag{3.49}$$

From the proof of Proposition 3.3, $[\eta, Ad_{b_-}L_1] \in \oplus_{i=-(n-1)}^{k'-1}\mathfrak{g}^{(i)}$, hence

$$(e_j, [\eta, Ad_{b_-}L_1]e_{k'+j}) = 0. \tag{3.50}$$

Therefore the above expression simplifies to

$$\begin{aligned}
x_{j,k'+j}(Ad_S^*(b,(\xi,\eta))^{-1}(L_0,(L_{-1},L_1)) =&(e_j, b_-L_0b_-^{-1}e_{k'+j})\\
=&\sum_{i,\ell}(b_-)_{ji}(L_0)_{i\ell}(b_-^{-1})_{\ell,k'+j}\\
=&\sum_{i=1}^{j}\sum_{\ell=k'+j}^{n}(b_-)_{ji}(b_-^{-1})_{\ell,k'+j}(e_i, L_0e_\ell).
\end{aligned} \tag{3.51}$$

Now $(e_i, L_0 e_\ell) = 0$ for $\ell - i \geq k' + 1$. On the other hand, the maximum value of $i - \ell$ in the above sum is equal to $n - k'$ which corresponds to $i = n - k' + j$ and $\ell = j$. Therefore the above sum reduces to $\frac{(b_-)_{jj}}{(b_-)_{k'+j,k'+j}} x_{j,k'+j}(L) = x_{j,k'+j}(L)$ as $b_- \in N_-$. This proves the assertion.

(d) In this case, we have

$$
\begin{aligned}
&z_{n-k'+j,j}(Ad_S^*(b,(\xi,\eta))^{-1}(L_0,(L_{-1},L_1)) = (e_{n-k'+j}, b_- L_1 b_-^{-1} e_j) \\
&= \sum_{i=1}^{n-k'+j} \sum_{\ell=j}^{n} (b_-)_{n-k'+j,i} (b_-^{-1})_{\ell j} (e_i, L_1 e_\ell).
\end{aligned} \tag{3.52}
$$

Now $(e_i, L_1 e_\ell) = 0$ for $i - \ell < n - k'$. On the other hand, the maximum value of $i - \ell$ is equal to $n - k'$, corresponding to $i = n - k' + j$ and $\ell = j$. Hence the above sum reduces to $\frac{(b_-)_{n-k'+j,n-k'+j}}{(b_-)_{jj}} z_{n-k'+j,j}(L) = z_{n-k'+j,j}(L)$, as required. □

From Proposition 3.4 (c) and (d) above, it follows that the entries on the k'-th superdiagaonal of L_0 and the $(n-k')$-th subdiagonal of L_1 will stay fixed under the coadjoint action of S. That is, the corresponding coordinate functions are coadjoint orbit invariants of S.

Let

$$
\begin{aligned}
P(L) &= \left(\prod_{j=1}^{n-k} x_{k+j,j} \prod_{j=1}^{k} y_{j,n-k+j} \right)(L), \\
P'(L) &= \left(\prod_{j=1}^{n-k'} x_{j,k'+j} \prod_{j=1}^{k'} z_{n-k'+j,j} \right)(L).
\end{aligned} \tag{3.53}
$$

Then the set

$$
\mathcal{G}' = \{ L \in \mathcal{T}^{(k,k')} \mid P(L), P'(L) \neq 0 \} \tag{3.54}
$$

is an open, dense subset of $\mathcal{T}^{(k,k')}$ which is invariant under the coadjoint action of S according to the above proposition.

We next introduce the following functions on $\mathcal{T}^{(k,k')}$:

(3.55)

$$
\begin{aligned}
\sigma_i &= \prod_{j=1}^{n_1-k_1} x_{k+i+(j-1)d,i+(j-1)d} \prod_{j=1}^{k_1} y_{i+(j-1)d,n-k+i+(j-1)d}, \quad i = 1, \ldots, d, \\
\sigma_i' &= \prod_{j=1}^{n_1'-k_1'} x_{i+(j-1)d',k'+i+(j-1)d'} \prod_{j=1}^{k_1'} z_{n-k'+i+(j-1)d',i+(j-1)d'}, \quad i = 1, \ldots, d'.
\end{aligned}
$$

Note that these functions correspond to the quantities introduced in [**vMM**, Section 1] which are used in defining the notion of *regular* periodic difference operators. For us, the meaning of these functions is given in the next result.

PROPOSITION 3.5. *The functions σ_i, $i = 1, \ldots, d$, σ_i', $i = 1, \ldots, d'$ are invariant under the coadjoint action of S.*

PROOF. The argument for the σ_i makes use of Proposition 3.4 (a), (b) above and is identical to the one in [**LN2**, Proposition 2.3]. The assertion for the σ_i' is clear from Proposition 3.4 (c), (d). □

In view of the above propositions, the set

$$\mathcal{G} = \{L \in \mathcal{G}' \mid \sigma_i(L) \neq \sigma_j(L), i \neq j, \sigma_i'(L) \neq \sigma_j'(L), i \neq j\} \tag{3.56}$$

is an open, dense subset of $\mathcal{T}^{(k,k')}$ invariant under the coadjoint action of S.

DEFINITION 3.6. A loop $L \in \mathcal{T}^{(k,k')}$ is generic iff $L \in \mathcal{G}$. An orbit $\mathcal{O}_L$ of the coadjoint action of $\widetilde{G}$ through $L \in \mathcal{T}^{(k,k')}$ is generic iff the loop L is generic.

We next introduce the characteristic polynomial of $L(z)$.

PROPOSITION 3.7. *For $L \in \mathcal{T}^{(k,k')}$, we have*

(a) $\det(L(z) - wI) = (-1)^n w^n + \sum_{r=0}^{n-1} \left(\sum_{j=-\lfloor \frac{k'd_r}{h} \rfloor}^{\lfloor \frac{kd_r}{h} \rfloor} I_{r,j}(L) z^{-j} \right) w^{n-1-r}$,

(b) $I_{r,\lfloor \frac{kd_r}{h} \rfloor}(L) \not\equiv 0$, $I_{r,-\lfloor \frac{k'd_r}{h} \rfloor}(L) \not\equiv 0$, $r = 1, \ldots, n-1$.

PROOF.

(a) For an $n \times n$ matrix $A \in \mathfrak{g}$, define the functions $E_r(A)$ by $\det(A - wI) = (-1)^n w^n + \sum_{r=0}^{n-1} E_{r+1}(A) w^{n-1-r}$. Then for each r, the expansion of $E_{r+1}(L(z))$ in z is a sum of terms of the form[1]

$$(L_0)_{i_1 j_1} \cdots (L_0)_{i_a j_a} (L_1)_{l_1 m_1} \cdots (L_1)_{l_b m_b} z^b (L_{-1})_{p_1 q_1} \cdots (L_{-1})_{p_c q_c} z^{-c}, \tag{3.57}$$

where

$$\begin{aligned} & a + b + c = d_r = r + 1. \\ & -k' \leq i_1 - j_1, \ldots, i_a - j_a \leq k, n - k' \leq l_1 - m_1, \ldots, l_b - m_b \leq n - 1, \\ & n - k \leq q_1 - p_1, \ldots, q_c - p_c \leq n - 1, \end{aligned} \tag{3.58}$$

and moreover,

$$(j_1 - i_1) + \cdots + (j_a - i_a) + (m_1 - l_1) + \cdots + (m_b - l_b) + (q_1 - p_1) + \cdots + (q_c - p_c) = 0. \tag{3.59}$$

Clearly, our goal is to get upper and lower bounds for $c - b$. To do so, first note that

$$\begin{aligned} (n - k)c &\leq (q_1 - p_1) + \cdots (q_c - p_c) \\ &= (i_1 - j_1) + \cdots + (i_a - j_a) + (l_1 - m_1) + \cdots + (l_b - m_b) \\ &\leq ka + (n - 1)b. \end{aligned} \tag{3.60}$$

Using this inequality, and the first line of (3.58), it follows that

$$\begin{aligned} n(c - b) &\leq k(a + c) - b \\ &\leq k(d_r - b) - b = kd_r - (k + 1)b \\ &\Longrightarrow c - b \leq \frac{kd_r}{h}. \end{aligned} \tag{3.61}$$

On the other hand, we have

$$\begin{aligned} -k'a + (n - k')b &\leq (i_1 - j_1) + \cdots + (i_a - j_a) + (l_1 - m_1) + \cdots + (l_b - m_b) \\ &= (q_1 - p_1) + \cdots + (q_c - p_c) \\ &\leq (n - 1)c \end{aligned} \tag{3.62}$$

[1]The numbers $m_1, \ldots, m_b$ which appear in the next equation are column indices which have nothing to do with the exponents of $\mathfrak{g} = \mathfrak{gl}(n, \mathbb{C})$

from which we find

$$(3.63)\qquad \begin{aligned} n(b-c) &\le k'(a+b)-c\\ &= k'(d_r-c)-c = k'd_r-(k+1)c,\\ &\Longrightarrow b-c \le \frac{k'd_r}{h}. \end{aligned}$$

Consequently, $-\lfloor\frac{k'd_r}{h}\rfloor \le c-b \le \lfloor\frac{kd_r}{h}\rfloor$ from which the assertion follows.
(b) To establish the assertion for $I_{r,\lfloor\frac{kd_r}{h}\rfloor}(L)$, observe that that the number $\lfloor\frac{kd_r}{h}\rfloor$ is either equal to $\lfloor\frac{km_r-1}{h}\rfloor$ or $1+\lfloor\frac{km_r-1}{h}\rfloor$. In the former case, the assertion is a consequence of our calculation in Theorem 5.12 as $\{\phi_{ab}, I_{r,\lfloor\frac{kd_r}{h}\rfloor}\}=1$ for $(a,b)=(r-\lfloor\frac{kd_r}{h}\rfloor, n-r)\in\Delta(I)^o\cap\mathbb{N}^2$ since $a=n-b-\lfloor\frac{k(h-b+1)}{h}\rfloor$. (See (5.15) below.) In the latter case, we will differentiate between the following two possibilities: either $\frac{kd_r}{h}\in\mathbb{N}$ or $\frac{kd_r}{h}\notin\mathbb{N}$. If $\frac{kd_r}{h}=a\in\mathbb{N}$, then clearly, $\lfloor\frac{kd_r}{h}\rfloor > \frac{km_r-1}{h}$ and so necessarily, $r\in E$ (see (3.65) below for the definition of E). In this case, we have $k_1d_r=an_1$ and as n_1 and k_1 are relatively prime, it follows that $n_1|d_r$. Hence there exists a positive integer j such that $d_r=n_1j$. Consequently, $\lfloor\frac{kd_r}{h}\rfloor=k_1j$ and as $I_{n_1j-1,k_1j}(L)=(-1)^{n-j}E_j(\sigma_1(L),\dots,\sigma_d(L))$, the assertion follows. Alternatively, we could look at terms in the expansion of $E_{r+1}(L(z))$ in z, i.e. terms of the form in (3.57), with $c-b=\lfloor\frac{kd_r}{h}\rfloor$. From the point of view of the estimates in part (a), the case $\frac{kd_r}{h}=a\in\mathbb{N}$ holds iff we have equalities in (3.60) and (3.61). Hence we arrive at the same conclusion, as before. In the second case where $\frac{kd_r}{h}\notin\mathbb{N}$, $I_{r,\lfloor\frac{kd_r}{h}\rfloor}(L)$ is again a sum of terms of the form in (3.57) where $c-b=\lfloor\frac{kd_r}{h}\rfloor$. Since we have

$$(3.64)\qquad c-b=\lfloor\frac{kd_r}{h}\rfloor<\frac{kd_r}{h},$$

that is, we have strict inequality in the third line of (3.61), it follows that we must have $b=0$ and so $c=\lfloor\frac{kd_r}{h}\rfloor$. Now from the proof in Proposition 2.11(b) of [**LN2**], we find that $r=\lfloor\frac{n_1\mu}{k_1}\rfloor$, where $\mu=1,\dots,k-1$. By a direct calculation, it is easy to check that for $r=\lfloor\frac{n_1\mu}{k_1}\rfloor$, the corresponding $c=\lfloor\frac{kd_r}{h}\rfloor$ is equal to μ. We can construct nontrivial terms in $I_{r,\lfloor\frac{kd_r}{h}\rfloor}$ in the following way. Take μ entries from the $(n-k)$-th superdiagonal of L_{-1} and form their product, then take $\lfloor\frac{(n-k)\mu}{k}\rfloor=a-1$ entries from the k-th subdiagonal of L_0 and form their product. (Note that it is possible that $a-1=0$.) Then in order to complete this to a term in $I_{r,\lfloor\frac{kd_r}{h}\rfloor}$ of the form in (3.57), all we need to do is to pick an entry from L_0 not lying on its k-th subdiagonal and multiply this entry with the two products formed previously. That this is so is due to the fact that the relation in (3.59) (with $b=0$) gives $(n-k)\mu=k\lfloor\frac{(n-k)\mu}{k}\rfloor+(i_a-j_a)$ from which we can show that $i_a-j_a<k$. The argument for $I_{r,-\lfloor\frac{k'd_r}{h}\rfloor}(L)$ is similar. □

Our next goal is to single out the coadjoint orbit invariants amongst the $I_{r,j}$. This is an important step towards counting the correct number of genuine integrals (i.e., ones which are not Casimirs) in involution. For this purpose, introduce

$$(3.65)\qquad \begin{aligned} E &= \left\{r \mid \lfloor\frac{kd_r}{h}\rfloor = 1+\lfloor\frac{km_r-1}{h}\rfloor\right\},\\ E' &= \left\{r \mid \lfloor\frac{k'd_r}{h}\rfloor = 1+\lfloor\frac{k'm_r-1}{h}\rfloor\right\}\setminus\{r\mid m_r=n_1's,\ s=1,\dots,d'-1\}. \end{aligned}$$

PROPOSITION 3.8.

(a) *E' can be characterized alternatively as*

$$E' = \left\{ r \mid \lfloor \frac{k'd_r}{h} \rfloor = 1 + \lfloor \frac{k'm_r}{h} \rfloor \right\}. \tag{3.66}$$

(b) $I_{r,\lfloor \frac{kd_r}{h} \rfloor}$ *is a Casimir on* $\mathcal{T}^{(k,k')}$ *if* $r \in E$ *or if* $r = 0$.

(c) $I_{r,-\lfloor \frac{k'd_r}{h} \rfloor}$ *is a Casmir on* $\mathcal{T}^{(k,k')}$ *if* $r \in E'$ *or if* $r = 0$.

(d) $\mathrm{Card}(E) = k + d - 1$, $\mathrm{Card}(E') = k'$.

PROOF.

(a) First of all, if $m_r = n_1' s$, a direct computation shows that $\lfloor \frac{k'd_r}{h} \rfloor = 1 + \lfloor \frac{k'm_r - 1}{h} \rfloor = k_1' s$, while $1 + \lfloor \frac{k'm_r}{h} \rfloor = 1 + k_1' s$. This shows

$$\left\{ r \mid \lfloor \frac{k'd_r}{h} \rfloor = 1 + \lfloor \frac{k'm_r}{h} \rfloor \right\} \subset E'. \tag{3.67}$$

Conversely, suppose $r \in E'$. We want to show $\lfloor \frac{k'm_r - 1}{h} \rfloor = \lfloor \frac{k'm_r}{h} \rfloor$. We will divide the argument into two cases.

CASE 1. $\frac{k'd_r}{h} \in \mathbb{N}$.

Let $\frac{k'd_r}{h} = a$, then $k_1' d_r = n_1' a$. Since $\gcd(n_1', k_1') = 1$, it follows that $n_1' | d_r$ and so there exists $b \in \mathbb{N}$ such that $d_r = n_1' b$. Therefore $m_r = n_1' b - 1$ and consequently $\lfloor \frac{k'm_r}{h} \rfloor = \lfloor \frac{k'(m_r - 1)}{h} \rfloor = k_1' b - 1$.

CASE 2. $\frac{k'd_r}{h} \notin \mathbb{N}$.

In this case, there exists a unique $c \in \mathbb{N}$ such that $\frac{k'm_r - 1}{h} < c < \frac{k'd_r}{h}$. Multiply both sides by h, we find $k'm_r - 1 < ch < k'd_r$. Since $k'd_r - (k'm_r - 1) = k' + 1$, it follows that there exists a unique positive integer $1 \le j \le k'$ such that $ch = k'd_r - j$. Thus

$$j = k'd_r - ch = d'(k_1' d_r - cn_1'). \tag{3.68}$$

In particular, this implies j must be a multiple of d'. Therefore, by using (3.68) together with the inequalties $0 < j \le k'$, we deduce that

$$n_1' \frac{c}{k_1'} - 1 < m_r \le n_1' \frac{c}{k_1'} \tag{3.69}$$

and so $m_r = \lfloor n_1' \frac{c}{k_1'} \rfloor$. From the same inequalities, we also obtain

$$\lfloor \frac{k'm_r}{h} \rfloor = c \text{ or } c - 1, \tag{3.70}$$

and the former holds iff $n_1' \frac{c}{k_1'} \in \mathbb{N}$ which would mean that $m_r = n_1' \frac{c}{k_1'}$. But since $r \in E'$, this cannot happen and therefore we must have

$$\lfloor \frac{k'm_r}{h} \rfloor = c - 1. \tag{3.71}$$

On the other hand, on using (3.68), we find

$$c - \frac{k' + 1}{h} < \frac{k'm_r - 1}{h} \le c - \frac{1}{h} \tag{3.72}$$

which also gives

$$\lfloor \frac{k'm_r - 1}{h} \rfloor = c - 1. \tag{3.73}$$

The assertion therefore follows from (3.71) and (3.73).
(b) and (c) For $X \in L\mathfrak{g}$, define

$$I_{r,j}(X) = \oint_{|z|=1} E_{r+1}(X(z))\, z^{j-1} \frac{dz}{2\pi \mathrm{i}}. \tag{3.74}$$

Then clearly, $I_{r,j}(L)$ is just the pullback of the ad-invariant function $I_{r,j}(X)$ to $\mathcal{T}^{(k,k')}$. To show that $I_{r,\lfloor \frac{kd_r}{h} \rfloor}$, $r \in E$; $I_{r,-\lfloor \frac{k'd_r}{h} \rfloor}$, $r \in E'$, are Casimirs, it suffices to show that $\{I_{r,\lfloor \frac{kd_r}{h} \rfloor}, F\}(L) = 0$, $\{I_{r,-\lfloor \frac{k'd_r}{h} \rfloor}, F\}(L) = 0$ for any smooth function F. For this purpose, let $\{E_{ii}\}_{i=1}^n$ be the usual basis of $\mathfrak{h}$ and x_i $i = 1, \ldots, n$ the corresponding coordinates. We will compute

$$\begin{aligned} dI_{r,j}(L)(z) &= z^j dE_{r+1}(L(z)) \\ &= z^j \left(\sum_{a \neq b} \frac{\partial E_{r+1}}{\partial x_{ba}}(L(z)) E_{ab} + \sum_{i=1}^{n-1} \frac{\partial E_{r+1}}{\partial x_i}(L(z)) E_{ii} \right) \end{aligned} \tag{3.75}$$

and then specialize to the case where $j = \lfloor \frac{kd_r}{h} \rfloor$ and $j = -\lfloor \frac{k'd_r}{h} \rfloor$.

From the definition of E_{r+1}, it is clear that $\frac{\partial E_{r+1}}{\partial x_{ba}}(L(z))$ is a Laurent polynomial in z $(x_{aa} = x_a)$ for each choice of a and b. We have the following claim:

(i) $\frac{\partial E_{r+1}}{\partial x_i}(L(z)) = \sum_{\lambda=-\lfloor \frac{km_r}{h} \rfloor}^{\lfloor \frac{k'm_r}{h} \rfloor} F_{r;i}^{\lambda} z^{\lambda}$,

(ii) $\frac{\partial E_{r+1}}{\partial x_{ba}}(L(z)) = \sum_{\lambda=-\lfloor \frac{kd_r}{h} \rfloor}^{\lfloor \frac{k'm_r-1}{h} \rfloor} F_{r;b,a}^{\lambda} z^{\lambda}$, $1 \leq b - a \leq k$,

(iii) $\frac{\partial E_{r+1}}{\partial x_{ba}}(L(z)) = \sum_{\lambda=-\lfloor \frac{km_r-1}{h} \rfloor}^{\lfloor \frac{k'm_r-1}{h} \rfloor} F_{r;b,a}^{\lambda} z^{\lambda}$, $k+1 \leq b - a \leq h - k' - 1$,

(iv) $\frac{\partial E_{r+1}}{\partial x_{ba}}(L(z)) = \sum_{\lambda=-1-\lfloor \frac{km_r-1}{h} \rfloor}^{\lfloor \frac{k'm_r-1}{h} \rfloor} F_{r;b,a}^{\lambda} z^{\lambda}$, $h - k' \leq b - a \leq h - 1$,

(v) $\frac{\partial E_{r+1}}{\partial x_{ba}}(L(z)) = \sum_{\lambda=-\lfloor \frac{km_r-1}{h} \rfloor}^{\lfloor \frac{k'd_r}{h} \rfloor} F_{r;b,a}^{\lambda} z^{\lambda}$, $1 \leq a - b \leq k'$,

(vi) $\frac{\partial E_{r+1}}{\partial x_{ba}}(L(z)) = \sum_{\lambda=-\lfloor \frac{km_r-1}{h} \rfloor}^{1+\lfloor \frac{k'm_r-1}{h} \rfloor} F_{r;b,a}^{\lambda} z^{\lambda}$, $k'+1 \leq a - b \leq h - k - 1$,

(vii) $\frac{\partial E_{r+1}}{\partial x_{ba}}(L(z)) = \sum_{\lambda=-\lfloor \frac{kd_r}{h} \rfloor+1}^{1+\lfloor \frac{k'm_r-1}{h} \rfloor} F_{r;b,a}^{\lambda} z^{\lambda}$, $h - k \leq a - b \leq h - 1$.

We will verify the above claim for (ii), (vi) and (vii), leaving the details of the others to the interested reader.

For (ii), note that $\frac{\partial E_{r+1}}{\partial x_{ba}}(L(z))$ is a linear combination of monomials of the form

$$\prod_{i-\ell=h-k'}^{h-1} (L_{-1}z)_{i\ell}^{\alpha(i,\ell)} (L_0)_{ba}^{\alpha(b,a)-1} \prod_{\substack{\ell-i=-k \\ (i,\ell)\neq(b,a)}}^{k'} (L_0)_{i\ell}^{\alpha(i,\ell)} \prod_{\ell-i=h-k}^{h-1} (L_{-1}z^{-1})_{i\ell}^{\alpha(i,\ell)}, \tag{3.76}$$

where $\alpha(b,a) \geq 1$ and

$$\begin{aligned} &\sum_{i-\ell=h-k'}^{h-1} \alpha(i,\ell) + \sum_{\substack{\ell-i=-k \\ (i,\ell)\neq(b,a)}}^{k'} \alpha(i,\ell) + \alpha(b,a) + \sum_{\ell-i=h-k}^{h-1} \alpha(i,\ell) = d_r, \\ &\sum_{i-\ell=h-k'}^{h-1} \alpha(i,\ell)(\ell-i) + \sum_{\substack{\ell-i=-k \\ (i,\ell)\neq(b,a)}}^{k'} \alpha(i,\ell)(\ell-i) + \alpha(b,a)(a-b) \\ &\quad + \sum_{\ell-i=h-k}^{h-1} \alpha(i,\ell)(\ell-i) = 0. \end{aligned} \tag{3.77}$$

By using the second relation in (3.77) above, we have the estimate

$$\begin{aligned} &(h-k) \sum_{\ell-i=h-k}^{h-1} \alpha(i,\ell) \\ &\leq \sum_{\ell-i=h-k}^{h-1} \alpha(i,\ell)(\ell-i) + \sum_{\ell-i=1}^{k'} \alpha(i,\ell)(\ell-i) \\ &= \sum_{i-\ell=h-k'}^{h-1} \alpha(i,\ell)(i-\ell) + \sum_{\substack{i-\ell=1 \\ (i,\ell)\neq(b,a)}}^{k} \alpha(i,\ell)(i-\ell) \\ &\quad + \alpha(b,a)(b-a) \\ &\leq (h-1) \sum_{i-\ell=h-k'}^{h-1} \alpha(i,\ell) + k \sum_{\substack{i-\ell=1 \\ (i,\ell)\neq(b,a)}}^{k} \alpha(i,\ell) + \alpha(b,a)(b-a). \end{aligned} \tag{3.78}$$

Therefore, when we take into account the first relation in (3.77), we find

$$\begin{aligned} &(h-k) \sum_{\ell-i=h-k}^{h-1} \alpha(i,\ell) \\ &\leq (h-1) \sum_{i-\ell=h-k'}^{h-1} \alpha(i,\ell) + k(d_r - \sum_{\ell-i=1}^{k'} \alpha(i,\ell) - \alpha(b,a) \\ &\quad - \sum_{i-\ell=h-k'}^{h-1} \alpha(i,\ell) - \sum_{\ell-i=h-k}^{h-1} \alpha(i,\ell)) + \alpha(b,a)(b-a) \\ &\leq (h-k-1) \sum_{i-\ell=h-k'}^{h-1} \alpha(i,\ell) - k \sum_{\ell-i=h-k}^{h-1} \alpha(i,\ell) + kd_r \\ &\quad - (k-(b-a))\alpha(b,a). \end{aligned} \tag{3.79}$$

Consequently,

$$\sum_{\ell-i=h-k}^{h-1} \alpha(i,\ell) \leq \left(1 - \frac{k+1}{h}\right) \sum_{i-\ell=h-k'}^{h-1} \alpha(i,\ell) + \frac{kd_r}{h} - \frac{k-(b-a)}{h}\alpha(b,a). \tag{3.80}$$

By using this inequality, we find that the exponent of z in (3.76) is

$$\begin{aligned}
&\sum_{i-\ell=h-k'}^{h-1} \alpha(i,\ell) - \sum_{\ell-i=h-k}^{h-1} \alpha(i,\ell) \\
&\geq \frac{k+1}{h} \sum_{i-\ell=h-k'}^{h-1} \alpha(i,\ell) - \frac{kd_r}{h} + \frac{k-(b-a))}{h}\alpha(b,a) \\
&\geq -\frac{kd_r}{h}.
\end{aligned} \tag{3.81}$$

In a similar way, by using the second relation in (3.77), we find

$$\begin{aligned}
&(h-k') \sum_{i-\ell=h-k'}^{h-1} \alpha(i,\ell) \\
&\leq \sum_{i-\ell=h-k'}^{h-1} \alpha(i,\ell)(i-\ell) + \sum_{\substack{i-\ell=1 \\ (i,\ell)\neq(b,a)}}^{k} \alpha(i,\ell)(i-\ell) \\
&= \sum_{\ell-i=h-k}^{h-1} \alpha(i,\ell)(\ell-i) + \sum_{\ell-i=1}^{k'} \alpha(i,\ell)(\ell-i) - \alpha(b,a)(b-a) \\
&\leq (h-1) \sum_{\ell-i=h-k}^{h-1} \alpha(i,\ell) + k' \sum_{\ell-i=1}^{k'} \alpha(i,\ell) - \alpha(b,a)(b-a).
\end{aligned} \tag{3.82}$$

Therefore, when we apply the first relation in (3.77) to the above estimate, we obtain

$$\begin{aligned}
&(h-k') \sum_{i-\ell=h-k'}^{h-1} \alpha(i,\ell) \\
&\leq (h-1) \sum_{\ell-i=h-k}^{h-1} \alpha(i,\ell) + k'(d_r - \sum_{\substack{i-\ell=1 \\ (i,\ell)\neq(b,a)}} \alpha(i,\ell) - \alpha(b,a) \\
&\quad - \sum_{i-\ell=h-k'}^{h-1} \alpha(i,\ell) - \sum_{\ell-i=h-k}^{h-1} \alpha(i,\ell)) - \alpha(b,a)(b-a) \\
&\leq (h-k'-1) \sum_{\ell-i=h-k}^{h-1} \alpha(i,\ell) - k' \sum_{i-\ell=h-k'}^{h-1} \alpha(i,\ell) + kd_r \\
&\quad - (k'+(b-a))\alpha(b,a).
\end{aligned} \tag{3.83}$$

This yields

$$\sum_{i-\ell=h-k'}^{h-1} \alpha(i,\ell) \leq \left(1 - \frac{k'+1}{h}\right) \sum_{\ell-i=h-k}^{h-1} \alpha(i,\ell) + \frac{kd_r}{h} - \frac{k'+(b-a))}{h}\alpha(b,a). \tag{3.84}$$

From this estimate, we find that the exponent of z in (3.76) is

$$\begin{aligned}
&\sum_{i-\ell=h-k'}^{h-1}\alpha(i,\ell)-\sum_{\ell-i=h-k}^{h-1}\alpha(i,\ell)\\
&\le -\frac{k'+1}{h}\sum_{\ell-i=h-k}^{h-1}\alpha(i,\ell)+\frac{k'd_r}{h}-\frac{k'+(b-a)}{h}\alpha(b,a)\\
&\le \frac{k'm_r-1}{h}.
\end{aligned}\tag{3.85}$$

The assertion in (ii) is now evident from (3.81) and (3.85).

For (vi), note that the partial derivative $\frac{\partial E_{r+1}}{\partial x_{ba}}(L(z))$ for $k'+1\le a-b\le h-k-1$ is a linear combination of monomials of the form

$$\prod_{i-\ell=h-k'}^{h-1}(L_1z)_{i\ell}^{\alpha(i,\ell)}\prod_{\ell-i=-k}^{k'}(L_0)_{i\ell}^{\alpha(i,\ell)}\prod_{\ell-i=h-k}^{h-1}(L_{-1}z^{-1})_{i\ell}^{\alpha(i,\ell)},\tag{3.86}$$

where in this case, we have

$$\begin{aligned}
&\sum_{i-\ell=h-k'}^{h-1}\alpha(i,\ell)+\sum_{\ell-i=-k}^{k'}\alpha(i,\ell)+\sum_{\ell-i=h-k}^{h-1}\alpha(i,\ell)+1=d_r,\\
&\sum_{i-\ell=h-k'}^{h-1}\alpha(i,\ell)(\ell-i)+\sum_{\ell-i=-k}^{k'}\alpha(i,\ell)(\ell-i)+\sum_{\ell-i=h-k}^{h-1}\alpha(i,\ell)(\ell-i)+(a-b)=0.
\end{aligned}\tag{3.87}$$

Using the second relation in (3.87), we have

$$\begin{aligned}
&(h-k)\sum_{\ell-i=h-k}^{h-1}\alpha(i,\ell)+(a-b)\\
&\le\sum_{\ell-i=h-k}^{h-1}\alpha(i,\ell)(\ell-i)+\sum_{\ell-i=1}^{k'}\alpha(i,\ell)(\ell-i)+(a-b)\\
&=\sum_{i-\ell=h-k'}^{h-1}\alpha(i,\ell)(i-\ell)+\sum_{i-\ell=1}^{k}\alpha(i,\ell)(i-\ell)\\
&\le(h-1)\sum_{i-\ell=h-k'}^{h-1}\alpha(i,\ell)+k\sum_{i-\ell=1}^{k}\alpha(i,\ell).
\end{aligned}\tag{3.88}$$

If we now apply the first relation in (3.87) to the above estimate, we find

$$
\begin{aligned}
&(h-k)\sum_{\ell-i=h-k}^{h-1}\alpha(i,\ell)+(a-b)\\
&\le (h-1)\sum_{i-\ell=h-k'}^{h-1}\alpha(i,\ell)+k(d_r-1-\sum_{i-\ell=h-k'}^{h-1}\alpha(i,\ell)\\
&\quad -\sum_{\ell-i=1}^{k'}\alpha(i,\ell)-\sum_{\ell-i=h-k}^{h-1}\alpha(i,\ell))\\
&\le (h-k-1)\sum_{i-\ell=h-k'}^{h-1}\alpha(i,\ell)+km_r-k\sum_{\ell-i=h-k}^{h-1}\alpha(i,\ell)
\end{aligned}
\tag{3.89}
$$

from which we obtain

$$
\sum_{\ell-i=h-k}^{h-1}\alpha(i,\ell)\le\left(1-\frac{k+1}{h}\right)\sum_{i-\ell=h-k'}^{h-1}\alpha(i,\ell)+\frac{km_r-(a-b)}{h}.
\tag{3.90}
$$

Hence it follows from this last inequality that the exponent of z in (3.86) is

$$
\begin{aligned}
&\sum_{i-\ell=h-k'}^{h-1}\alpha(i,\ell)-\sum_{\ell-i=h-k}^{h-1}\alpha(i,\ell)\\
&\ge\frac{k+1}{h}\sum_{i-\ell=h-k'}^{h-1}\alpha(i,\ell)-\frac{km_r-(a-b)}{h}\\
&\ge\frac{-km_r+(a-b)}{h}\ge\frac{-km_r+(k'+1)}{h}\\
&>\frac{-km_r+1}{h}.
\end{aligned}
\tag{3.91}
$$

To get an upper estimate of the exponent of z, we proceed in a similar way, using (3.87):

(3.92)

$$\begin{aligned}(h-k')\sum_{i-\ell=h-k'}^{h-1}\alpha(i,\ell) &\leq \sum_{i-\ell=h-k'}^{h-1}\alpha(i,\ell)(i-\ell)+\sum_{i-\ell=1}^{k}\alpha(i,\ell)(i-\ell)\\ &= \sum_{\ell-i=h-k}^{h-1}\alpha(i,\ell)(\ell-i)+\sum_{\ell-i=1}^{k'}\alpha(i,\ell)(\ell-i)+(a-b)\\ &\leq (h-1)\sum_{\ell-i=h-k}^{h-1}\alpha(i,\ell)+k'\sum_{\ell-i=1}^{k'}\alpha(i,\ell)+(a-b)\\ &= (h-1)\sum_{\ell-i=h-k}^{h-1}\alpha(i,\ell)+k'(d_r-1-\sum_{i-\ell=1}^{k}\alpha(i,\ell)\\ &\quad -\sum_{i-\ell=h-k'}^{h-1}\alpha(i,\ell)-\sum_{\ell-i=h-k}^{h-1}\alpha(i,\ell))+(a-b)\\ &\leq (h-k'-1)\sum_{\ell-i=h-k}^{h-1}\alpha(i,\ell)+k'm_r-k'\sum_{\ell-i=h-k'}^{h-1}\alpha(i,\ell)\\ &\quad +(a-b).\end{aligned}$$

This gives

$$\sum_{i-\ell=h-k'}^{h-1}\alpha(i,\ell)\leq\left(1-\frac{k'+1}{h}\right)\sum_{\ell-i=h-k}^{h-1}\alpha(i,\ell)+\frac{k'm_r+(a-b)}{h}. \tag{3.93}$$

From this estimate, we find that the exponent in z in (3.86) is

$$\begin{aligned}&\sum_{i-\ell=h-k'}^{h-1}\alpha(i,\ell)-\sum_{\ell-i=h-k}^{h-1}\alpha(i,\ell)\\ &\leq -\frac{k'+1}{h}\sum_{\ell-i=h-k}^{h-1}\alpha(i,\ell)+\frac{k'm_r+(a-b)}{h}\\ &\leq -\frac{k'+1}{h}\sum_{\ell-i=h-k}^{h-1}\alpha(i,\ell)+\frac{k'm_r+h-k-1}{h}\\ &= \frac{k'm_r-1}{h}+1-\frac{k}{h}-\frac{k'+1}{h}\sum_{\ell-i=h-k}^{h-1}\alpha(i,\ell)\\ &\leq 1+\frac{k'm_r-1}{h}.\end{aligned} \tag{3.94}$$

Combining the inequalities in (3.91) and (3.94), the assertion in (vi) follows.

For (vii), we begin with the fact that $\frac{\partial E_{r+1}}{\partial x_{ba}}(L(z))$, $h-k \le a-b \le h-1$ is a linear combination of monomials of the form

$$\prod_{i-\ell=h-k'}^{h-1} (L_1 z)_{i\ell}^{\alpha(i,\ell)} \prod_{\ell-i=-k}^{k'} (L_0)_{i\ell}^{\alpha(i,\ell)} (L_{-1}z^{-1})_{ba}^{\alpha(b,a)-1} \prod_{\substack{\ell-i=h-k \\ (i,\ell)\neq(b,a)}}^{h-1} (L_{-1}z^{-1})_{i\ell}^{\alpha(i,\ell)}, \tag{3.95}$$

where $\alpha(b,a) \ge 1$ and in this case, we have the following relations:

$$\begin{gathered} \sum_{i-\ell=h-k'}^{h-1} \alpha(i,\ell) + \sum_{\ell-i=-k}^{k'} \alpha(i,\ell) + \sum_{\substack{\ell-i=h-k \\ (i,\ell)\neq(b,a)}}^{h-1} \alpha(i,\ell) + \alpha(b,a) = d_r, \\ \sum_{i-\ell=h-k'}^{h-1} \alpha(i,\ell)(\ell-i) + \sum_{\ell-i=-k}^{k'} \alpha(i,\ell)(\ell-i) + \sum_{\substack{\ell-i=h-k \\ (i,\ell)\neq(b,a)}}^{h-1} \alpha(i,\ell)(\ell-i) \\ + \alpha(b,a)(a-b) = 0. \end{gathered} \tag{3.96}$$

Proceeding as before, it follows on using (3.96) that

$$\begin{aligned} &\alpha(b,a)(a-b) + (h-k) \sum_{\substack{\ell-i=h-k \\ (i,\ell)\neq(b,a)}}^{h-1} \alpha(i,\ell) \\ &\le \sum_{\substack{\ell-i=h-k \\ (i,\ell)\neq(b,a)}}^{h-1} \alpha(i,\ell)(\ell-i) + \alpha(b,a)(a-b) + \sum_{\ell-i=1}^{k'} \alpha(i,\ell)(\ell-i) \\ &= \sum_{i-\ell=h-k'}^{h-1} \alpha(i,\ell)(i-\ell) + \sum_{i-\ell=1}^{k} \alpha(i,\ell)(i-\ell) \\ &\le (h-1) \sum_{i-\ell=h-k'}^{h-1} \alpha(i,\ell) + k \sum_{i-\ell=1}^{k} \alpha(i,\ell) \\ &= (h-1) \sum_{i-\ell=h-k'}^{h-1} \alpha(i,\ell) + k(d_r - \sum_{i-\ell=h-k'}^{h-1} \alpha(i,\ell) - \sum_{\ell-i=1}^{k'} \alpha(i,\ell) - \sum_{i-\ell=h-k}^{h-1} \alpha(i,\ell)) \\ &\le (h-k-1) \sum_{i-\ell=h-k'}^{h-1} \alpha(i,\ell) + kd_r - k \sum_{\substack{i-\ell=h-k \\ (i,\ell)\neq(b,a)}}^{h-1} \alpha(i,\ell) - k\alpha(b,a). \end{aligned} \tag{3.97}$$

Therefore,

$$h \sum_{\substack{\ell-i=h-k \\ (i,\ell)\neq(b,a)}}^{h-1} \alpha(i,\ell) \le (h-k-1) \sum_{i-\ell=h-k'}^{h-1} \alpha(i,\ell) + kd_r - (k+(a-b))\alpha(b,a) \tag{3.98}$$

from which we obtain

(3.99)

$$\sum_{\ell-i=h-k}^{h-1} \alpha(i,\ell) \leq \left(1-\frac{k+1}{h}\right) \sum_{i-\ell=h-k'}^{h-1} \alpha(i,\ell) + \frac{kd_r}{h} + \frac{h-k-(a+b)}{h}\alpha(b,a).$$

From this inequality, we find that the exponent of z in (3.95) is

$$\begin{aligned} &\sum_{i-\ell=h-k'}^{h-1} \alpha(i,\ell) - \sum_{\substack{\ell-i=h-k \\ (i,\ell)\neq(b,a)}}^{h-1} \alpha(i,\ell) - \alpha(b,a) + 1 \\ &\geq \frac{k+1}{h} \sum_{i-\ell=h-k'}^{h-1} \alpha(i,\ell) - \frac{kd_r}{h} + 1 + \frac{(a+b)-(h-k)}{h}\alpha(b,a) \\ &\geq -\frac{kd_r}{h} + 1. \end{aligned} \tag{3.100}$$

On the other hand, using the second relation in (3.96) gives

$$\begin{aligned} &(h-k') \sum_{i-\ell=h-k'}^{h-1} \alpha(i,\ell) \\ &\leq \sum_{i-\ell=h-k'}^{h-1} \alpha(i,\ell)(i-\ell) + \sum_{i-\ell=1}^{k} \alpha(i,\ell)(i-\ell) \\ &= \sum_{\substack{\ell-i=h-k \\ (i,\ell)\neq(b,a)}}^{h-1} \alpha(i,\ell)(\ell-i) + \alpha(b,a)(a-b) + \sum_{\ell-i=1}^{k'} \alpha(i,\ell)(\ell-i) \\ &\leq (h-1) \sum_{\substack{\ell-i=h-k \\ (i,\ell)\neq(b,a)}}^{h-1} \alpha(i,\ell) + \alpha(b,a)(a-b) + k' \sum_{\ell-i=1}^{k'} \alpha(i,\ell) \end{aligned} \tag{3.101}$$

Therefore when we apply the first relation in (3.96) to the above estimate, we find

(3.102)

$$\begin{aligned} &(h-k') \sum_{i-\ell=h-k'}^{h-1} \alpha(i,\ell) \\ &\leq (h-1) \sum_{\substack{\ell-i=h-k \\ (i,\ell)\neq(b,a)}}^{h-1} \alpha(i,\ell) + \alpha(b,a)(a-b) + k'(d_r - \sum_{i-\ell=h-k'}^{h-1} \alpha(i,\ell) \\ &\quad - \sum_{i-\ell=1}^{k} \alpha(i,\ell) - \sum_{\substack{i-\ell=h-k \\ (i,\ell)\neq(b,a)}}^{h-1} \alpha(i,\ell) - \alpha(b,a)) \\ &\leq (h-k'-1) \sum_{\substack{i-\ell=h-k \\ (i,\ell)\neq(b,a)}}^{h-1} \alpha(i,\ell) + (a-b-k')\alpha(b,a) - k' \sum_{i-\ell=h-k'}^{h-1} \alpha(i,\ell) + k'd_r. \end{aligned}$$

This gives

$$h\sum_{i-\ell=h-k'}^{h-1}\alpha(i,\ell)\le (h-k'-1)\sum_{\substack{\ell-i=h-k\\(i,\ell)\neq(b,a)}}^{h-1}\alpha(i,\ell)+(a-b-k')\alpha(b,a)+k'd_r \tag{3.103}$$

from which we obtain the estimate

$$\sum_{i-\ell=h-k'}^{h-1}\alpha(i,\ell)\le\left(1-\frac{k'+1}{h}\right)\sum_{\substack{\ell-i=h-k\\(i,\ell)\neq(b,a)}}^{h-1}\alpha(i,\ell)+\frac{k'd_r}{h}+\frac{a-b-k'}{h}\alpha(b,a). \tag{3.104}$$

On using (3.104), we find that the exponent of z in (3.95) is

$$\begin{aligned}
&\sum_{i-\ell=h-k'}^{h-1}\alpha(i,\ell)-\sum_{\substack{\ell-i=h-k\\(i,\ell)\neq(b,a)}}^{h-1}\alpha(i,\ell)-\alpha(b,a)+1\\
&\le -\frac{k'+1}{h}\sum_{\substack{\ell-i=h-k\\(i,\ell)\neq(b,a)}}^{h-1}\alpha(i,\ell)+\frac{k'd_r}{h}\alpha(b,a)+1\\
&\le \frac{k'd_r}{h}+1-\frac{h-(a-b)+k'}{h}\alpha(b,a)\\
&\le \frac{k'm_r-1}{h}+1.
\end{aligned} \tag{3.105}$$

Hence the assertion in (vii) follows from (3.100) and (3.105)

Thus it follows from the above relations (i)-(vii) that

$$\begin{aligned}
&dI_{r,j}(L)(z)\\
&=\sum_{\lambda=j-\lfloor\frac{km_r}{h}\rfloor}^{j+\lfloor\frac{k'm_r}{h}\rfloor}\left(\sum_{i=1}^{n-1}F_{r;i}^{\lambda-j}E_{ii}\right)z^{\lambda}+\sum_{\lambda=j-\lfloor\frac{kd_r}{h}\rfloor}^{j+\lfloor\frac{k'm_r-1}{h}\rfloor}\left(\sum_{b-a=1}^{k}F_{r;b,a}^{\lambda-j}E_{ab}\right)z^{\lambda}\\
&+\sum_{\lambda=j-\lfloor\frac{km_r-1}{h}\rfloor}^{j+\lfloor\frac{k'm_r-1}{h}\rfloor}\left(\sum_{b-a=k+1}^{h-k'-1}F_{r;b,a}^{\lambda-j}E_{ab}\right)z^{\lambda}\\
&+\sum_{\lambda=j-1-\lfloor\frac{km_r-1}{h}\rfloor}^{j+\lfloor\frac{k'm_r-1}{h}\rfloor}\left(\sum_{b-a=h-k'}^{h-1}F_{r;b,a}^{\lambda-j}E_{ab}\right)z^{\lambda}\\
&+\sum_{\lambda=j-\lfloor\frac{km_r-1}{h}\rfloor}^{j+\lfloor\frac{k'd_r}{h}\rfloor}\left(\sum_{a-b=1}^{k'}F_{r;b,a}^{\lambda-j}E_{ab}\right)z^{\lambda}+\sum_{j-\lfloor\frac{km_r-1}{h}\rfloor}^{j+1+\lfloor\frac{k'm_r-1}{h}\rfloor}\left(\sum_{a-b=k'+1}^{h-k-1}F_{r;b,a}^{\lambda-j}E_{ab}\right)z^{\lambda}\\
&+\sum_{\lambda=j-\lfloor\frac{kd_r}{h}\rfloor+1}^{j+\lfloor\frac{k'm_r-1}{h}\rfloor+1}\left(\sum_{a-b=h-k}^{h-1}F_{r;b,a}^{\lambda-j}E_{ab}\right)z^{\lambda}.
\end{aligned} \tag{3.106}$$

If $r\in E$, $j=\lfloor\frac{kd_r}{h}\rfloor$, $dI_{r,\lfloor\frac{kd_r}{h}\rfloor}(L)$ is a polynomial in z with coefficients in $\mathfrak{g}$ (this is not the case if $\lfloor\frac{kd_r}{h}\rfloor=\lfloor\frac{km_r-1}{h}\rfloor$). Since the constant term of this polynomial is given by

$$\sum_{i=1}^{n-1} F_{r;i}^{-\lfloor\frac{kd_r}{h}\rfloor} E_{ii} + \sum_{b-a=1}^{k} F_{r;b,a}^{-\lfloor\frac{kd_r}{h}\rfloor} E_{ab} + \sum_{b-a=h-k'}^{h-1} F_{r;b,a}^{-\lfloor\frac{kd_r}{h}\rfloor} E_{ab}, \tag{3.107}$$

it follows from this expression that $dI_{r,\lfloor\frac{kd_r}{h}\rfloor}(L) \in \widetilde{\mathfrak{b}}_+$ and so $\Pi_{\widetilde{\mathfrak{n}}_-} dI_{r,\lfloor\frac{kd_r}{h}\rfloor}(L) = 0$. Consequently, for any smooth function F on $\widetilde{\mathfrak{g}}$, we have

$$\begin{aligned}
\{I_{r,\lfloor\frac{kd_r}{h}\rfloor}, F\}(L) &= (L, [\Pi_{\widetilde{\mathfrak{b}}_+} dI_{r,\lfloor\frac{kd_r}{h}\rfloor}(L), \Pi_{\widetilde{\mathfrak{b}}_+} dF(L)])_{L\mathfrak{g}} \\
&\quad - (L, [\Pi_{\widetilde{\mathfrak{n}}_-} dI_{r,\lfloor\frac{kd_r}{h}\rfloor}(L), \Pi_{\widetilde{\mathfrak{n}}_-} dF(L)])_{L\mathfrak{g}} \\
&= ([L, dI_{r,\lfloor\frac{kd_r}{h}\rfloor}(L)], \Pi_{\widetilde{\mathfrak{b}}_+} dF(L))_{L\mathfrak{g}} \\
&= 0.
\end{aligned} \tag{3.108}$$

If $r \in E'$, $j = -\lfloor\frac{k'd_r}{h}\rfloor$, it follows from (3.106) that $dI_{r,-\lfloor\frac{k'd_r}{h}\rfloor}(L)$ is a polynomial in z^{-1} with coefficients in $\mathfrak{g}$ and the constant term in this polynomial is given by

$$\sum_{a-b=1}^{h-1} F_{r;b,a}^{\lfloor\frac{k'd_r}{h}\rfloor} E_{ab} \in \mathfrak{n}_-. \tag{3.109}$$

Hence $dI_{r,-\lfloor\frac{k'd_r}{h}\rfloor}(L) \in \widetilde{\mathfrak{n}}_-$ and so for any smooth function F on $\widetilde{\mathfrak{g}}$, we have

$$\begin{aligned}
\{I_{r,-\lfloor\frac{k'd_r}{h}\rfloor}, F\}(L) &= -(L, [\,dI_{r,-\lfloor\frac{k'd_r}{h}\rfloor}(L), \Pi_{\widetilde{\mathfrak{n}}_-} dF(L)\,])_{L\mathfrak{g}} \\
&= -([L, dI_{r,-\lfloor\frac{k'd_r}{h}\rfloor}(L)], \Pi_{\widetilde{\mathfrak{n}}_-} dF(L))_{L\mathfrak{g}} \\
&= 0.
\end{aligned} \tag{3.110}$$

Finally, it is clear from (3.41) that $\operatorname{tr} L_0$ (equivalently, $I_{00}(L)$) is an invariant of the coadjoint action of S. This completes the proof of (b) and (c).
(b) In [**LN2**, Proposition 2.11 (b)], we have derived the formula for Card(E). The formula for Card(E') is now clear. □

REMARK 3.9. In Proposition 3.8 above, we have picked out the coadjoint orbit invariants amongst the $I_{r,j}$. However, it is important to point out that there are coadjoint orbit invariants of $\widetilde{G}$ in $\mathcal{G}$ which are not provided by the coefficients of the polynomial $\det(L(z) - wI)$ in (z, z^{-1}, w). Indeed, examples of such invariants already appeared in Proposition 3.4 (c) and (d), where we showed that the entries on the k'-th superdiagonal of L_0 and the $(n-k')$-th subdiagonal of L_1 are fixed under the coadjoint action of $\widetilde{G}$, hence the corresponding coordinate functions are coadjoint orbit invariants. In the next section, we will compute the dimension of the generic coadjoint orbits in $\mathcal{G}$ using an approach which is quite different from the one used in [**LN2**].

CHAPTER 4

Dimension of Generic Coadjoint Orbits

Recall that

$$G_J = \{b_+ b_-^{-1} \mid b_+ \in B_+, b_- \in N_-\} \tag{4.1}$$

with multiplication defined by

$$b \cdot b' = b_+ b' b_-^{-1} \tag{4.2}$$

is a Lie group that integrates $\mathfrak{g}_J = \mathfrak{b}_+ \ominus \mathfrak{n}_-$. The group G_J acts on $\mathfrak{g} \oplus \mathfrak{g}$ via the formula

$$\rho(b)(\xi, \eta) = (Ad_{b_+}\xi, Ad_{b_-}\eta). \tag{4.3}$$

Hence we have the semidirect product Lie group $S = G_J \ltimes_\rho (\mathfrak{g} \oplus \mathfrak{g})$ with multiplication given by

$$(b, (\xi, \eta))(b', (\xi', \eta')) = (b \cdot b', (\xi, \eta) + \rho(b)(\xi', \eta')). \tag{4.4}$$

To begin with, we will rewrite the expression (3.41) for the coadjoint action of the S on $\mathfrak{s}^* \simeq \mathfrak{g} \times (\mathfrak{g} \oplus \mathfrak{g})$ in the form

$$\begin{aligned} &Ad_S^*(b, (\xi, \eta))^{-1}(L_0, (L_{-1}, L_1)) \\ &= (\, \Pi_{\mathfrak{n}_+}(Ad_{b_-}\Pi_{\mathfrak{n}_+} L_0 + [Ad_{b_-}L_1, \eta]) + \Pi_{\mathfrak{b}_-}(Ad_{b_+}\Pi_{\mathfrak{b}_-}L_0 + [\xi, Ad_{b_+}L_{-1}]), \\ &\quad (Ad_{b_+}L_{-1}, Ad_{b_-}L_1)\,), \end{aligned} \tag{4.5}$$

and recall that $\mathcal{G} \subset \mathcal{T}^{(k,k')}$ is invariant under this action. We want to compute the dimension of the generic coadjoint orbits in $\mathcal{G}$. To this end, note that B_+ acts on $\mathfrak{g}$ via $b_+ \cdot X = Ad_{b_+}X$ and we have the semidirect product group $B_+ \ltimes \mathfrak{g}$ with multiplication

$$(b_+, X_1)(b_+', X_2) = (b_+ b_+', X_1 + Ad_{b_+}X_2). \tag{4.6}$$

A direct calculation shows that the coadjoint action of $B_+ \ltimes \mathfrak{g}$ on the dual of its Lie algebra, identified with $\mathfrak{b}_- \times \mathfrak{g}$, is given by

$$\begin{aligned} &Ad^*_{B_+ \ltimes \mathfrak{g}}(b_+, X)^{-1}(\Pi_{\mathfrak{b}_-}L_0, L_{-1}) \\ =&(\Pi_{\mathfrak{b}_-}(Ad_{b_+}\Pi_{\mathfrak{b}_-}L_0 + [X, Ad_{b_+}L_{-1}]), Ad_{b_+}L_{-1}). \end{aligned} \tag{4.7}$$

On the other hand, if we take N_- with multiplication which it inherits from G_J, then N_- acts on $\mathfrak{g}$ via $b_-^{-1} \cdot Y = Ad_{b_-}Y$ and we have the corresponding semidirect product group $N_- \ltimes \mathfrak{g}$. Identifying the dual of its Lie algebra with $\mathfrak{n}_+ \times \mathfrak{g}$, we find that the coadjoint action of $N_- \ltimes \mathfrak{g}$ is given by

$$\begin{aligned} &Ad^*_{N_- \ltimes \mathfrak{g}}(b_-^{-1}, Y)^{-1}(\Pi_{\mathfrak{n}_+}L_0, L_1) \\ =&(\Pi_{\mathfrak{n}_+}(Ad_{b_-}\Pi_{\mathfrak{n}_+}L_0 + [Ad_{b_-}L_1, Y]), Ad_{b_-}L_1). \end{aligned} \tag{4.8}$$

On comparing (4.7) and (4.8) with (4.5), we see that we can identify the coadjoint orbit of S through $(L_0, (L_{-1}, L_1))$ with the product of the $B_+ \ltimes \mathfrak{g}$-coadjoint orbit through $(\Pi_{\mathfrak{b}_-} L_0, L_{-1})$ and the $N_- \ltimes \mathfrak{g}$-coadjoint orbit through $(\Pi_{\mathfrak{n}_+} L_0, L_1)$. Consequently, the problem of computing the dimension of Ad^*_S-orbits in $\mathcal{G}$ decouples and it suffices to compute the dimension of the $B_+ \ltimes \mathfrak{g}$ orbit through generic $(\Pi_{\mathfrak{b}_-} L_0, L_{-1})$ as well as those of the $N_- \ltimes \mathfrak{g}$ orbit through generic $(\Pi_{\mathfrak{n}_+} L_0, L_1)$.

We will take $L_0 = L_0'$, $L_1 = L_1'$, $L_{-1} = L_{-1}'$, where

$$
\begin{aligned}
L_0' &= \sum_{i=1}^{n-k'} a_{i,k'+i} E_{i,k'+i} + \sum_{i=1}^{n-k} a_{k+i,i} E_{k+i,i}, \\
L_1' &= \sum_{i=1}^{k'} a_{n-k'+i,i} E_{n-k'+i,i}, \qquad (4.9)\\
L_{-1}' &= \sum_{i=1}^{k} a_{i,n-k+i} E_{i,n-k+i}
\end{aligned}
$$

satisfy the genericity assumptions, that is, the corresponding $L' \in \mathcal{G}$.

We have

$$
\begin{aligned}
&\text{dimension of the } N_- \ltimes \mathfrak{g} \text{ coadjoint orbit through } (\Pi_{\mathfrak{n}_+} L_0', L_1') \\
&= \dim \mathfrak{n}_- \times \mathfrak{g} - \dim(\mathfrak{n}_- \times \mathfrak{g})_{(\Pi_{\mathfrak{n}_+} L_0', L_1')},
\end{aligned} \qquad (4.10)
$$

where $(\mathfrak{n}_- \times \mathfrak{g})_{(\Pi_{\mathfrak{n}_+} L_0', L_1')}$ is the isotropy subalgebra through $(\Pi_{\mathfrak{n}_+} L_0', L_1')$, given by

$$
\begin{aligned}
&(\mathfrak{n}_- \times \mathfrak{g})_{(\Pi_{\mathfrak{n}_+} L_0', L_1')} \\
&= \{(B, Y) \in \mathfrak{n}_- \times \mathfrak{g} \mid \Pi_{\mathfrak{n}_+}([B, \Pi_{\mathfrak{n}_+} L_0'] + [Y, L_1']) = 0, [B, L_1'] = 0\},
\end{aligned} \qquad (4.11)
$$

where for convenience, we have replaced Y by $-Y$. Since

$$\Pi_{\mathfrak{n}_+} L_0' \in \mathfrak{g}^{(k')}, \Pi_{\mathfrak{n}_+}[B, \Pi_{\mathfrak{n}_+} L_0]$$

depend on $B^{(-j)}$ only for $0 < j < k'$. Similarly, since $L_1' \in \mathfrak{g}^{(-(n-k'))}$, it follows that $\Pi_{\mathfrak{n}_-}[Y, L_1']$ depend on $Y^{(j)}$ only for $n - k' < j \le n-1$. The notation used in the previous two lines is the following: for $X \in \mathfrak{g}$, we write $X = \sum_{j=-(n-1)}^{n-1} X^{(j)}$, where $X^{(j)} \in \mathfrak{g}^{(j)}$ (see (1.9) for the definition of $\mathfrak{g}^{(j)}$). We will use this notation except in the subsection below where we need to use block matrices.

Since the isotropy subalgebra is defined by a system of homogeneous linear equations, the variables which are not present in the equations are automatically free variables. By counting, the number of such variables is given by

$$
\begin{aligned}
&[1 + \cdots + (n-1)] - [(n-1) + \cdots + (n - (k'-1))] - [1 + \cdots + (k'-1)] \\
&+ n^2 \\
&= [1 + \cdots + (n-1)] + n^2 - n(k'-1).
\end{aligned} \qquad (4.12)
$$

We now consider in detail the linear system which defines the isotropy subalgebra:

$$
\begin{aligned}
&\Pi_{\mathfrak{n}_+}([B, \Pi_{\mathfrak{n}_+} L_0'] + [Y, L_1']) = 0, \\
&[B, L_1'] = 0.
\end{aligned} \qquad (4.13)
$$

To do so, we will partition our matrices into $d' \times d'$ blocks. So let us begin with some new notation. If $M \in \mathfrak{g}$, we will regard M as a partitioned matrix $(M_{ij})_{i,j=1}^{n_1'}$, where each M_{ij} is $d' \times d'$. Since the $d' \times d'$ submatrix M_{ij} lies in block row i and block column j of the matrix M, we will call it the (i,j) block of M. For $j > 0$, we let $\mathfrak{g}_{d'}^{(j)}$ (resp. $\mathfrak{g}_{d'}^{(-j)}$) be the set of matrices in $\mathfrak{g}$ partitioned into $d' \times d'$ blocks whose entries outside the j-th block superdiagonal (resp. subdiagonal) are zeros. Analogously, for $j=0$, we denote by $\mathfrak{g}_{d'}^{(0)}$ the set of matrices $M \in \mathfrak{g}$ which are block diagonal with $d' \times d'$ diagonal blocks M_{ii}, $i=1,\ldots,n_1'$. Clearly, we have the direct sum decomposition

$$\mathfrak{g} = \oplus_{j\in\mathbb{Z}}\, \mathfrak{g}_{d'}^{(j)}, \tag{4.14}$$

where $\mathfrak{g}_{d'}^{(j)} = \mathfrak{g}_{d'}^{(-j)} = 0$ for $j > n_1'-1$. Thus if $M \in \mathfrak{g}$, we will write

$$M = \sum_{j=-(n_1'-1)}^{n_1'-1} M^{(j)}, \quad M^{(j)} \in \mathfrak{g}_{d'}^{(j)}. \tag{4.15}$$

Note that the notation $M^{(j)}$ was used with a different meaning previously, but in the following we will use it with this new meaning, the two meanings of course are identical in the case when $d'=1$. Using this notation, the linear system in (4.13) can be rewritten as

$$\begin{aligned} &\sum_{j=0}^{k_1'-1} [B^{(-j)}, \Pi_{\mathfrak{n}_+} L_0'] + \Pi_{\mathfrak{n}_+}([B^{(-k_1')}, \Pi_{\mathfrak{n}_+} L_0']) \\ &+ \sum_{j=n_1'-k_1'+1}^{n_1'-1} [Y^{(j)}, L_1'] + \Pi_{\mathfrak{n}_+}[Y^{(n_1'-k_1')}, L_1'] = 0, \\ &\sum_{j=0}^{k_1'-1} ([B^{(-j)}, L_1']) = 0. \end{aligned} \tag{4.16}$$

Since $[B^{(-j)}, \Pi_{\mathfrak{n}_+} L_0'] \in \mathfrak{g}_{d'}^{(k_1'-j)}$, and $[Y^{(j)}, L_1'] \in \mathfrak{g}_{d'}^{(j-(n_1'-k_1'))}$, the above equations are equivalent to

$$\begin{aligned} &[B^{(-j)}, \Pi_{\mathfrak{n}_+} L_0'] + [Y^{(n_1'-j)}, L_1'] = 0, \quad j=0,\ldots,k_1'-1, \\ &\Pi_{\mathfrak{n}_+}([B^{(-k_1')}, \Pi_{\mathfrak{n}_+} L_0'] + [Y^{(n_1'-k_1')}, L_1']) = 0, \\ &[B^{(-j)}, L_1'] = 0, \quad j=0,\ldots,k_1'-1, \end{aligned} \tag{4.17}$$

where $Y^{(n_1')} = 0$.

In order to represent our matrices in partitioned form and to do calculations efficiently with them, we introduce for $1 \le i \le n_1'$ the $n \times d'$ matrix $E_{(i)}, 1 \le i \le n_1'$, partitioned into $d' \times d'$ blocks with its i-th block equal to the $d' \times d'$ identity matrix $I_{d'}$ and zeros elsewhere. For a matrix H which is $d' \times d'$, note that $E_{(i)} H E_{(j)}^T \in \mathfrak{g}$ is the partitioned matrix whose (i,j) block is H and has zeros elsewhere. Thus we can write

(4.18)

$$\begin{aligned} B^{(-j)} &= \sum_{i=0}^{n_1'-j} E_{(i+j)} B_i^{(-j)} E_{(i)}^T, \quad \Pi_{\mathfrak{n}_+} L_0' = \sum_{i=1}^{n_1'-k_1'} E_{(i)} A_i^{(k_1')} E_{(k_1'+i)}^T, \\ Y^{(n_1'-j)} &= \sum_{i=1}^{j} E_{(i)} Y_i^{(n_1'-j)} E_{(n_1'-j+i)}^T, \quad L_1' = \sum_{i=1}^{k_1'} E_{(n_1'-k_1'+i)} A_i^{(-(n_1'-k_1'))} E_{(i)}^T. \end{aligned}$$

Using $E_{(i)}^T E_{(j)} = \delta_{ij} I_{d'}$, a direct calculation shows that

$$
\begin{aligned}
&[B^{(-j)}, \Pi_{\mathfrak{n}_+} L_0'] + [Y^{(n_1'-j)}, L_1'] \\
&= \sum_{i=1}^{j} E_{(i)} Y_i^{(n_1'-j)} A_{k_1'-j+i}^{(-(n_1'-k_1'))} E_{(k_1'-j+i)}^T + \sum_{i=j+1}^{n_1'-k_1'+j} E_{(i)} B_{i-j}^{(-j)} A_{i-j}^{(k_1')} E_{(k_1'-j+i)}^T \\
&\quad - \sum_{i=1}^{n_1'-k_1'} E_{(i)} A_i^{(k_1')} B_{k_1'-j+i}^{(-j)} E_{(k_1'-j+i)}^T \\
&\quad - \sum_{i=n_1'-k_1'+1}^{n_1'-k_1'+j} E_{(i)} A_{i-(n_1'-k_1')}^{(-(n_1'-k_1'))} Y_{i-(n_1'-k_1')}^{(n_1'-j)} E_{(k_1'-j+i)}^T,
\end{aligned} \tag{4.19}
$$

and

$$
\begin{aligned}
&[B^{(-j)}, L_1'] \\
&= \sum_{i=1}^{k_1'-j} E_{(n_1'-k_1'+i+j)} \left(B_{n_1'-k_1'+i}^{(-j)} A_i^{(-(n_1'-k_1'))} - A_{i+j}^{(-(n_1'-k_1'))} B_i^{(-j)} \right) E_{(i)}^T.
\end{aligned} \tag{4.20}
$$

In particular, when $j = k_1'$ in (4.19), we obtain

$$
\begin{aligned}
&[B^{(-k_1')}, \Pi_{\mathfrak{n}_+} L_0'] + [Y^{(n_1'-k_1')}, L_1'] \\
&= \sum_{i=1}^{k_1'} E_{(i)} Y_i^{(n_1'-k_1')} A_i^{(-(n_1'-k_1'))} E_{(i)}^T + \sum_{i=k_1'+1}^{n_1'} E_{(i)} B_{i-k_1'}^{(-k_1')} A_{i-k_1'}^{(k_1')} E_{(i)}^T \\
&\quad - \sum_{i=1}^{n_1'-k_1'} E_{(i)} A_i^{(k_1')} B_i^{(-k_1')} E_{(i)}^T - \sum_{i=n_1'-k_1'+1}^{n_1'} E_{(i)} A_{i-(n_1'-k_1')}^{(-(n_1'-k_1'))} Y_{i-(n_1'-k_1')}^{(n_1'-k_1')} E_{(i)}^T.
\end{aligned} \tag{4.21}
$$

Now define M_i for all $i \in \mathbb{Z}$ by

$$
\begin{aligned}
&M_i = A_i^{(k_1')}, \quad i = 1, \ldots, n_1' - k_1', \\
&M_{n_1'-k_1'+i} = A_i^{(-(n_1'-k_1'))}, \quad i = 1, \ldots, k_1', \\
&M_{i+n_1'} = M_i, \quad \text{for all } i \in \mathbb{Z}.
\end{aligned} \tag{4.22}
$$

Similarly, define $S_i^{(n_1'-j)}$ for all $i \in \mathbb{Z}$ in terms of $Y_i^{(n_1'-j)}$ and $B_i^{(-j)}$ via the formula:

$$
\begin{aligned}
&S_i^{(n_1'-j)} = Y_i^{(n_1'-j)}, \quad i = 1, \ldots, j, \\
&S_{j+i}^{(n_1'-j)} = B_i^{(-j)}, \quad i = 1, \ldots, n_1' - j, \\
&S_{i+n_1'}^{(n_1'-j)} = S_i^{(n_1'-j)}, \quad \text{for all } i \in \mathbb{Z}.
\end{aligned} \tag{4.23}
$$

Then from (4.19),

$$
\begin{aligned}
&[B^{(-j)}, \Pi_{\mathfrak{n}_+} L_0'] + [Y^{(n_1'-j)}, L_1'] = 0 \\
&\Longleftrightarrow \sum_{i=1}^{n_1'-k_1'+j} E_{(i)} \left(S_i^{(n_1'-j)} M_{i-j} - M_i S_{i+k_1'}^{(n_1'-j)} \right) E_{(k_1'-j+i)}^T = 0, \\
&j = 0, 1, \ldots, k_1' - 1.
\end{aligned} \tag{4.24}
$$

Similarly, it follows from (4.20) that

(4.25)

$$\begin{aligned}
&[B^{(-j)}, L_1'] = 0\\
\iff &\sum_{i=1}^{k_1'-j} E_{(n_1'-k_1'+i+j)} \left(S^{(n_1'-j)}_{n_1'-k_1'+j+i} M_{n_1'-k_1'+i} - M_{n_1'-k_1'+i+j} S^{(n_1'-j)}_{i+j} \right) E^T_{(i)} = 0,\\
\iff &\sum_{i=-(k_1'-j)+1}^{0} E_{(n_1'+i)} (S_i^{(n_1'-j)} M_{i-j} - M_i S^{(n_1'-j)}_{i+k_1'}) E^T_{(i+k_1'-j)} = 0,\\
&j = 0, 1, \ldots, k_1' - 1.
\end{aligned}$$

Lastly, if for a $d' \times d'$ matrix H, $\Pi_{\mathfrak{n}_+(d')} H$ denotes the projection to the strictly upper triangular part of H, it follows from (4.21) that we have

(4.26)

$$\begin{aligned}
&\Pi_{\mathfrak{n}_+}([B^{(-k_1')}, \Pi_{\mathfrak{n}_+} L_0'] + [Y^{(n_1'-k_1')}, L_1']) = 0\\
\iff &\sum_{i=1}^{n_1'} E_{(i)} \left(\left(\Pi_{\mathfrak{n}_+(d')} S_i^{(n_1'-k_1')} \right) M_{n_1'-k_1'+i} - M_i \left(\Pi_{\mathfrak{n}_+(d')} S^{(n_1'-k_1')}_{k_1'+i} \right) \right) E^T_{(i)} = 0.
\end{aligned}$$

Combining (4.24) and (4.25), we conclude for $j = 0, 1, \ldots, k_1' - 1$ that

$$S^{(n_1'-j)}_{i+k_1'} = M_i^{-1} S_i^{(n_1'-j)} M_{i-j}, \quad i \in \mathbb{Z}. \tag{4.27}$$

On the other hand, (4.26) gives

$$\Pi_{\mathfrak{n}_+(d')}\, S^{(n_1'-k_1')}_{i+k_1'} = M_i^{-1} \left(\Pi_{\mathfrak{n}_+(d')}\, S^{(n_1'-k_1')} \right) M_{i-k_1'}. \tag{4.28}$$

PROPOSITION 4.1.

(a) *For $j = 1, \ldots, k_1' - 1$, we can solve for $S_i^{(n_1'-j)}$, $i = 2, \ldots, n_1'$ in terms of $S_1^{(n_1'-j)}$, which is diagonal. On the other hand, $S_i^{(n_1')} = 0$ for all i.*

(b) *$\Pi_{\mathfrak{n}_+(d')} S_i^{(n_1'-k_1')} = 0$ for all i.*

PROOF.

(a) By iterating the relation in (4.27), we find

$$S^{(n_1'-j)}_{1+ik_1'} = \left(\prod_{a=0}^{i-1} M_{1+ak_1'} \right)^{-1} S_1^{(n_1'-j)} \left(\prod_{a=0}^{i-1} M_{1-j+ak_1'} \right) \tag{4.29}$$

for all i. Since $1 + ik_1' \pmod{n_1'}$, $i = 1, \ldots, n_1' - 1$ are all distinct, the first part of our assertion follows. To show that $S_1^{(n_1'-j)}$ is diagonal, we again iterate the relation in (4.27); this gives

$$S_1^{(n_1'-j)} = \left(\prod_{i=0}^{n_1'-1} M_{1+in_1'} \right)^{-1} S_1^{(n_1'-j)} \left(\prod_{i=0}^{n_1'-1} M_{1-j+ik_1'} \right). \tag{4.30}$$

However, since the integers $1-j+i+k_1' \pmod{n_1'}$, $i = 0, \ldots, n_1'-1$ are all distinct, we conclude that

$$\prod_{i=0}^{n_1'-1} M_{1+in_1'} = \prod_{i=0}^{n_1'-1} M_{1-j+ik_1'}. \tag{4.31}$$

Therefore

$$S_1^{(n_1'-j)} = \left(\prod_{i=0}^{n_1'-1} M_{1+in_1'}\right)^{-1} S_1^{(n_1'-j)} \left(\prod_{i=0}^{n_1'-1} M_{1+in_1'}\right). \tag{4.32}$$

But by the definition of the $\sigma_i'(L)$, we have

$$\prod_{i=0}^{n_1'-1} M_{1+in_1'} = \operatorname{diag}(\sigma_1'(L), \ldots, \sigma_{d'}'(L)) \tag{4.33}$$

and as $\sigma_i'(L) \neq \sigma_j'(L)$, $i \neq j$, it follows from (4.32) that $S_1^{(n_1'-j)}$ is diagonal, $j = 1$, $\ldots$, $k_1' - 1$. When $j = 0$, $S_1^{(n_1')} = B_1^{(0)}$ is strictly lower triangular since $B \in \mathfrak{n}_-$. Hence $S_1^{(n_1')} = 0$ and this implies $S_i^{(n_1')} = 0$ for all i.
(b) The argument is similar to the one we used in part (a). □

COROLLARY 4.2. *The dimension of the $N_- \ltimes \mathfrak{g}$ coadjoint orbit through $(\Pi_{\mathfrak{n}_+} L_0'$, $L_1')$ is given by $n(k'-1) - (k'-d')$.*

PROOF. The above proposition shows that we have an additional $(k_1'-1)d' = k' - d'$ free variables for the linear system in (4.13). Therefore, by (4.10), (4.12) and the discussion above, we find

$$\begin{aligned}
&\text{dimension of the } N_- \ltimes \mathfrak{g} \text{ coadjoint orbit through } (\Pi_{\mathfrak{n}_+} L_0', L_1') \\
&= \dim \mathfrak{n}_- \times \mathfrak{g} - \dim(\mathfrak{n}_- \times \mathfrak{g})_{(\Pi_{\mathfrak{n}_+} L_0', L_1')} \\
&= [1 + \cdots + (n-1)] + n^2 - [1 + \cdots + (n-1)] - n^2 + n(k_1'-1) - (k'-d') \\
&= n(k'-1) - (k'-d').
\end{aligned} \tag{4.34}$$

□

We now dispense with our usage of partitioned matrices since there is no further use of it in the rest of the section. In the proof of our next result, we will switch back to the usual notation. In particular, for $X \in \mathfrak{g}$, recall that we have the decomposition $X = \sum_{j=-(n-1)}^{n-1} X^{(j)}$, where $X^{(j)} \in \mathfrak{g}^{(j)}$, $-(n-1) \leq j \leq n-1$.

THEOREM 4.3. *The dimension of the Ad_S^*-orbits in $\mathcal{G} \subset \mathcal{T}^{(k,k')}$ is given by $(n-1)(k+k') + d' - d$.*

PROOF. In view of the above corollary and our earlier discussion, it suffices to show that the dimension of the $B_+ \ltimes \mathfrak{g}$ coadjoint orbit through $(\Pi_{\mathfrak{b}_-} L_0', L_{-1}')$ is given by $(n-1)k + n - d$. Analogous to the other case, we have

$$\begin{aligned}
&\text{dimension of the } B_+ \ltimes \mathfrak{g} \text{ coadjoint orbit through } (\Pi_{\mathfrak{b}_-} L_0', L_{-1}') \\
&= \dim \mathfrak{b}_+ \times \mathfrak{g} - \dim(\mathfrak{b}_+ \times \mathfrak{g})_{(\Pi_{\mathfrak{b}_-} L_0', L_{-1}')}
\end{aligned} \tag{4.35}$$

where the isotropy subalgebra $(\mathfrak{b}_+ \times \mathfrak{g})_{(\Pi_{\mathfrak{b}_-} L_0', L_{-1}')}$ through $(\Pi_{\mathfrak{b}_-} L_0', L_{-1}')$ is given by

$$\begin{aligned}
&(\mathfrak{b}_+ \times \mathfrak{g})_{(\Pi_{\mathfrak{b}_-} L_0', L_{-1}')} \\
&= \{(U, X) \in \mathfrak{b}_+ \times \mathfrak{g} \mid \Pi_{\mathfrak{b}_-}([U, \Pi_{\mathfrak{n}_-} L_0'] + [X, L_{-1}']) = 0, [U, L_{-1}'] = 0\}.
\end{aligned} \tag{4.36}$$

We will rewrite the equations that define the isotropy subalgebra. First of all,

$$[U, L_{-1}'] = 0 \Longleftrightarrow [\Pi_{\mathfrak{n}_+} U, L_{-1}'] = 0, \ [\operatorname{diag} U, L_{-1}'] = 0. \tag{4.37}$$

Next, by the definition of $\Pi_{\mathfrak{b}_-}$, we have

(4.38)
$$\begin{aligned}&\Pi_{\mathfrak{b}_-}([U,\Pi_{\mathfrak{n}_-}L_0']+[X,L_{-1}'])\\ =&\Pi_{\mathfrak{b}_-}([\operatorname{diag}U,\Pi_{\mathfrak{n}_-}L_0']+[\Pi_{\mathfrak{n}_+}U,\Pi_{\mathfrak{n}_-}L_0']+[X,L_{-1}'])\\ =&\Pi_{\mathfrak{n}_-}([\Pi_{\mathfrak{n}_+}U,\Pi_{\mathfrak{n}_-}L_0']+[X,L_{-1}'])+\operatorname{diag}([U^{(k)},\Pi_{\mathfrak{n}_-}L_0']+[X^{(-(n-k))},L_{-1}'])\\ &+\left[\operatorname{diag}U,\Pi_{\mathfrak{n}_-}L_0'\right].\end{aligned}$$

Since $\Pi_{\mathfrak{n}_-}([\Pi_{\mathfrak{n}_+}U,\Pi_{\mathfrak{n}_-}L_0']+[X,L_{-1}'])\in\oplus_{j=1}^{k-1}\mathfrak{g}^{(-j)}$, $[\operatorname{diag}U,\Pi_{\mathfrak{n}_-}L_0']\in\mathfrak{g}^{(-k)}$, it follows from the above decomposition that

$$\begin{aligned}&\Pi_{\mathfrak{b}_-}([U,\Pi_{\mathfrak{n}_-}L_0']+[X,L_{-1}'])=0\\ \Longleftrightarrow&\Pi_{\mathfrak{n}_-}([\Pi_{\mathfrak{n}_+}U,\Pi_{\mathfrak{n}_-}L_0']+[X,L_{-1}'])=0,\\ &[\operatorname{diag}U,\Pi_{\mathfrak{n}_-}L_0']=0,\\ &\operatorname{diag}(([U^{(k)},\Pi_{\mathfrak{n}_-}L_0']+[X^{(-(n-k))},L_{-1}'])=0.\end{aligned}\tag{4.39}$$

Clearly, the two equations

$$\Pi_{\mathfrak{n}_-}([\Pi_{\mathfrak{n}_+}U,\Pi_{\mathfrak{n}_-}L_0']+[X,L_{-1}'])=0,\quad [\Pi_{\mathfrak{n}_+}U,L_{-1}']=0\tag{4.40}$$

are the analogs of the ones which define the isotropy subalgebra $(\mathfrak{n}_-\times\mathfrak{g})_{(\Pi_{\mathfrak{n}_+}L_0',L_1')}$. From the method of our analysis above, it is clear that the number of additional free variables from this linear system is given by $k-d$. On the other hand, the equations involving $\operatorname{diag}U$ in (4.37)and (4.39) can be merged into one single equation

$$[\operatorname{diag}U,\Pi_{\mathfrak{n}_-}L_0'+L_{-1}]=0\tag{4.41}$$

from which we find that

$$\begin{aligned}&u_{ii}-u_{i+k,i+k}=0,\quad 1\le i\le n-k',\\ &u_{ii}-u_{n-k+i,n-k+i}=0,\quad 1\le i\le k'.\end{aligned}\tag{4.42}$$

Therefore if we extend u_{ii} to all $i\in\mathbb{Z}$ by the periodicity condition $u_{i+n,i+n}=u_{ii}$, we infer that

$$u_{i+d,i+d}=u_{ii}\quad\text{for all } i\in\mathbb{Z}.\tag{4.43}$$

In other words, the linear system (4.41) has d free variables. We now turn to the last equation in (4.39). Explicitly, we have

$$\begin{aligned}&\sum_{i=1}^{n-k}a_{i+k,i}u_{i,i+k}E_{ii}+\sum_{i=1}^{k}a_{i,n-k+i}x_{n-k+i,i}E_{n-k+i,n-k+i}\\ &-\sum_{i=1}^{n-k}a_{i+k,i}u_{i,i+k}E_{i+k,i+k}-\sum_{i=1}^{k}a_{i,n-k+i}x_{n-k+i,i}E_{ii}=0.\end{aligned}\tag{4.44}$$

Since $a_{ij}=a_{\alpha,\beta}$ when $i\equiv\alpha\pmod n$, $j\equiv\beta\pmod n$, if we define s_i, $i\in\mathbb{Z}$ such that

$$\begin{aligned}&s_i=u_{i,i+k},\quad i=1,\dots,n-k;\ s_{n-k+i}=x_{n-k+i,i},\quad i=1,\dots,k,\\ &s_{i+n}=s_i\quad\text{for all } i\in\mathbb{Z},\end{aligned}\tag{4.45}$$

then (4.44) gives

$$a_{i+k,i}s_i-a_{i,n-k+i}s_{n-k+i}=0\quad\text{for all } i\in\mathbb{Z},\tag{4.46}$$

or equivalently,

$$a_{i+2k,i+k}s_{i+k} = a_{i+k,i}s_i \quad \text{for all } i \in \mathbb{Z}. \tag{4.47}$$

From this, we see that from s_i, $1 \leq i \leq d$, we can recursively obtain s_{i+jd}, $1 \leq j \leq n_1 - 1$. Since the n positive integers $i + jd \pmod{n}$, $i = 1, \ldots, d$; $j = 0, \ldots, n_1 - 1$ are all distinct, we conclude that $s_1, \ldots, s_d$ are free variables for (4.46). Now $\Pi_{\mathfrak{b}_-}[U, \Pi_{\mathfrak{n}_-} L_0']$ depends on $U^{(j)}$ only for $j \leq k$. Similarly, $\Pi_{\mathfrak{b}_-}[X, L_{-1}']$ depends on $X^{(-j)}$ only for $n - k \leq j \leq n - 1$. Therefore, the number of variables which are not present in the equation is given by

$$(1 + \cdots + n) - [(n-k) + \cdots + n] + n^2 - (1 + \cdots + k). \tag{4.48}$$

Consequently,

(4.49)

$$\begin{aligned}
&\text{dimension of the } B_+ \ltimes \mathfrak{g} \text{ coadjoint orbit through } (\Pi_{\mathfrak{b}_-} L_0', L_{-1}') \\
&= \dim \mathfrak{b}_+ \times \mathfrak{g} - \dim(\mathfrak{b}_+ \times \mathfrak{g})_{(\Pi_{\mathfrak{b}_-} L_0', L_{-1}')} \\
&= (1 + \cdots + n) + n^2 - (1 + \cdots + n) + [(n-k) + \cdots + n] - n^2 + (1 + \cdots + k) \\
&\quad - (k - d) - d - d \\
&= [(n-k) + \cdots + n] + (1 + \cdots + k) - k - d \\
&= n(k+1) - k - d = (n-1)k + n - d,
\end{aligned}$$

as required. □

CHAPTER 5

Action-Angle Variables

PROPOSITION 5.1.
(a) *The Hamiltonian of the flow defined by* (1.10) *is given by*

$$H(L) = \frac{1}{2}(L,L)_{L\mathfrak{g}} = \frac{(-1)^{n-1}}{2} I_{10}(L). \tag{5.1}$$

(b) *The functions* $I_{r,j}(L)$, $-\lfloor\frac{k'd_r}{h}\rfloor + 1 \le j \le \lfloor\frac{kd_r}{h}\rfloor - 1$, $1 \le r \le n-1$, $I_{r,\lfloor\frac{kd_r}{h}\rfloor}$, $r \notin E \cup \{0\}$, $I_{r,-\lfloor\frac{k'd_r}{h}\rfloor}$, $r \notin E' \cup \{0\}$, *provide a collection of nontrivial integrals in involution on the generic coadjoint orbits in* $\mathcal{T}^{(k,k')}$. *Moreover, the number of functions in this collection is given by*

$$\frac{1}{2}[(n-1)(k+k') + d' - d]. \tag{5.2}$$

PROOF.
(a) Let $E_r(A)$ be the functions introduced in the proof of Proposition 3.7. Then we have the relation

$$(\operatorname{tr} A)^2 = \operatorname{tr} A^2 + (-1)^{n-2} E_2(A) \tag{5.3}$$

from which we obtain the formula for $H(L)$.
(b) The $I_{r,j}$ are the pullback of ad-invariant functions on $L\mathfrak{g} \simeq \widetilde{\mathfrak{g}}^*$ to the Poisson submanifold $\mathcal{T}^{(k,k')}$. Since ad-invariant functions on $\widetilde{\mathfrak{g}}$ Poisson commute by the abstract involution theorem in [**STS**], it follows that we have a collection of Poisson commuting integrals on the generic coadjoint orbits in $\mathcal{T}^{(k,k')}$. To count the number of nontrivial integrals in this collection, note that for $-\lfloor\frac{k'd_r}{h}\rfloor \le j \le \lfloor\frac{kd_r}{h}\rfloor, 1 \le r \le n-1$,

$$\begin{aligned}
&\#\text{ of } I_{r,j}\\
&= \sum_{r=1}^{n-1}\left(\lfloor\frac{kd_r}{h}\rfloor + 1 + \lfloor\frac{k'd_r}{h}\rfloor\right)\\
&= \frac{(n-1)(k-1)}{2} + \frac{d-1}{2} + k + \frac{(n-1)(k'-1)}{2} + \frac{d'-1}{2} + k' - (n-1)\\
&= \frac{(n-1)(k+k')}{2} + \frac{d+d'-2}{2} + k + k'
\end{aligned} \tag{5.4}$$

where we have used the formula in [**Kn**, Exercise 37, page 42] in going from the second line to the third line above. On the other hand, it follows from Proposition 3.8 that

$$\operatorname{Card}(E) = k + d - 1, \quad \operatorname{Card}(E') = k'. \tag{5.5}$$

On subtracting these two numbers from (5.4), the assertion follows. □

From Proposition 5.1 (b) and Theorem 4.3, we thus see that the number of nontrivial integrals in involution is precisely one half the dimension of the generic coadjoint orbits. We will establish the functional independence of these integrals on an open dense subset of $\mathcal{G}$ via Poisson bracket relations between these quantities and an associated set of (essentially) angle variables, to be constructed below.

For $L \in \mathcal{T}^{(k,k')}$, we introduce the polynomial

$$\begin{aligned} I(L,z,w) &= \det(zL(z) - wI) \\ &= (-1)^n w^n + \sum_{r=0}^{n-1} \left(\sum_{j=-\lfloor \frac{k'd_r}{h} \rfloor}^{\lfloor \frac{kd_r}{h} \rfloor} I_{r,j}(L) z^{r+1-j} \right) w^{n-1-r}. \end{aligned} \tag{5.6}$$

Then the equation $I(L,z,w) = 0$ defines an affine curve $C_{\mathrm{aff}}(L)$. Because $zL(z)$ is a matrix loop of degree 2 in z, we will compactify $C_{\mathrm{aff}}(L)$ in the weighted projective space $\mathbb{P}^2(1,1,2) = (\mathbb{C}^3 \setminus \{0\})/\mathbb{C}^*$, where the action of $\mathbb{C}^*$ on $\mathbb{C}^3 \setminus \{0\}$ is given by

$$\lambda \cdot (\zeta, z, w) = (\lambda\zeta, \lambda z, \lambda^2 w). \tag{5.7}$$

Thus we embed $C_{\mathrm{aff}}(L)$ in $\mathbb{P}^2(1,1,2)$ via the map

$$\mathbb{C}^2 \longrightarrow \mathbb{P}^2(1,1,2), \quad (z,w) \mapsto [1 : z : w], \tag{5.8}$$

we will denote its closure in $\mathbb{P}^2(1,1,2)$ by $C(L)$. Clearly, $C(L)$ is defined by the equation

$$\det(L_1 z^2 + L_0 z\zeta + L_{-1}\zeta^2 - wI) = 0 \tag{5.9}$$

and it meets the line at infinity at the point $P_\infty = [0 : 1 : 0]$. An easy calculation then shows that P_∞ is a singular point of $C(L)$. Since the complex compact surface $\mathbb{P}^2(1,1,2)$ has a singularity at $[0:0:1]$ and $[0:0:1] \notin C(L)$, it follows that $C(L)$ is embedded in the smooth complex surface $\mathbb{P}^2(1,1,2) \setminus \{[0:0:1]\}$, which can be identified with the total space of the line bundle $\mathcal{O}_{\mathbb{P}^1}(2)$ (cf. [**Aud**]).

We will study the properties of the spectral curve $C(L)$ and its normalization via the Newton polygon of $I(L,z,w)$, which turns out to be quite simple. To do so, recall that for a polynomial $f(z,w) = \sum_{\mu,\nu} a_{\mu,\nu} z^\mu w^\nu$, the carrier (support) of f is the set

$$\mathrm{carr}(f) = \{(\mu,\nu) \mid a_{\mu,\nu} \neq 0\}. \tag{5.10}$$

The convex hull of $\mathrm{carr}(f)$ is called the Newton polygon of f and will be denoted by $\Delta(f)$. We say that f is non-degenerate with respect to $\Delta(f)$ if for every face $\tau \subset \Delta(f)$ (including $\Delta(f)$ itself), the system

$$f_\tau = \frac{\partial f_\tau}{\partial z} = \frac{\partial f_\tau}{\partial w} = 0 \tag{5.11}$$

has no solution in $(\mathbb{C}^*)^2$ and f_τ is the polynomial $\sum_{(\mu,\nu)\in\tau} a_{\mu,\nu} z^\mu w^\nu$. Our main tool in studying $C(L)$ and its normalization is the following theorem.

THEOREM 5.2 ([**B, Kho**]). *Let $f(z,w) = 0$ define an irreducible curve C and let $\phi : \widetilde{C} \longrightarrow C$ be the normalization of C. If f is nondegenerate with respect to the Newton polygon $\Delta(f)$ and the singular points of the corresponding projective curve are among $[0:0:1]$, $[0:1:0]$, and $[1:0:0]$, then*

(a) *the genus of the non-singular model* $\widetilde{C}$ *of* C *is given by*

$$g(\widetilde{C}) = \mathrm{Card}(\Delta(f)^o \cap \mathbb{N}^2), \tag{5.12}$$

where $\Delta(f)^o$ *is the interior of* $\Delta(f)$,
(b) *if we denote the pullback of the coordinate functions* z *and* w *on* C *by* ϕ *also by the same symbols, then a basis of holomorphic* 1*-forms of* $\widetilde{C}$ *is given by*

$$\omega_{ab} = \frac{z^{a-1}w^{b-1}}{\partial f/\partial w}\, dz,\ (a,b) \in \Delta(f)^o \cap \mathbb{N}^2. \tag{5.13}$$

REMARK 5.3. The holomorphic 1-forms on $\widetilde{C}$ can also be computed from the formula [**BK**]

$$\frac{p(z,w)}{\partial f/\partial w}\, dz, \tag{5.14}$$

where $p(z,w) = 0$ is the equation of the curve of order $\deg f - 3$ adjoint to C. Thus the advantage of using the Baker-Khovanskii theorem in our present case is that it saves us from having to compute the adjoint polynomials [**N**].

PROPOSITION 5.4. *The Newton polygon* $\Delta(I)$ *of the spectral curve* $C(L)$ *is the triangular region with vertices at* $(0,n)$, $(n-k,0)$ *and* $(n+k',0)$.

PROOF. Change the index of summation in (5.6) by putting $a = r - j + 1$, $b = n - 1 - r$. Then $j = n - a - b$ and $r = n - 1 - b$. Since $0 \le r \le n-1$, we have $0 \le b \le n-1$. Then for given b in this range, we find

$$n - b - \lfloor \frac{k(n-b)}{h} \rfloor \le a \le n - b + \lfloor \frac{k'(n-b)}{h} \rfloor. \tag{5.15}$$

Therefore we can rewrite (5.6) in the form

$$I(L,z,w) = (-1)w^n + \sum_{b=0}^{n-1} \left(\sum_{a=n-b-\lfloor \frac{k(n-b)}{h} \rfloor}^{n-b+\lfloor \frac{k'(n-b)}{h} \rfloor} I_{n-1-b,n-a-b} z^a \right) w^b. \tag{5.16}$$

From this expression, it follows that if $(a,b) \in \mathrm{carr}(I)$, then

$$n - b - \frac{k}{n}(n-b) \le a \le n - b + \frac{k'(n-b)}{n}, \tag{5.17}$$

from which we obtain

$$-a\frac{n}{n-k} \le b - n \le -a\frac{n}{n+k'}. \tag{5.18}$$

This shows $\mathrm{carr}(I) \subset \Delta$, where Δ is the triangular region in the ab-plane bounded by the lines $b = 0$, $b - n = -\frac{n}{n-k}a$ and $b - n = -\frac{n}{n+k'}a$. On the other hand, it is clear from the expression for $I(L,z,w)$ above that the vertices $(n+k',0)$, $(0,n)$ and $(n-k,0)$ of $\partial\Delta$ are in $\mathrm{carr}(I)$. Consequently, $\Delta = \Delta(I)$. □

PROPOSITION 5.5.
(a) *The point* P_0 *with coordinates* $[\zeta : z : w] = [1:0:0]$ *is a non-ordinary singular point of order* $n-k$ *of* $C(L)$.
(b) *The point* $P_\infty = [0:1:0]$ *is a non-ordinary singular point of order* $n - k'$.
(c) *For* $L \in \mathcal{G}$ *satisfying additional genericity assumptions, the curve* $C(L)$ *is irreducible and has singular points only at* P_0 *and* P_∞.

PROOF.

(a) From the previous proposition, the polynomial $I(L,z,w)$ is a sum of terms of the form $z^a w^b$, where

$$n - \lfloor \frac{k(n-b)}{n} \rfloor \le a+b \le n + \lfloor \frac{k'(n-b)}{n} \rfloor. \tag{5.19}$$

From this, we see that the minimum value of $a+b$ is given by $n-k$, corresponding to $a=n-k$ and $b=0$. Since the corresponding term in $I(z,w)$ is $I_{n-1,k}(L)z^{n-k}$ and as $I_{n-1,k}(L) \neq 0$, it follows that $P_0 = [1:0:0]$ is a non-ordinary singular point of order $n-k$.

(b) In a neighborhood of P_∞ in $\mathbb{P}^2(1,1,2)$, introduce the coordinates

$$z' = \frac{1}{z}, \quad w' = \frac{w}{z^2}, \tag{5.20}$$

then the equation $I(L,z,w)=0$ gives

$$\begin{aligned} 0 &= \det(L_1 + L_0 z' + L_{-1}(z')^2 - w'I) \\ &= (-1)^n (w')^n + \sum_{r=0}^{n-1} \left(\sum_{j=-\lfloor \frac{k' d_r}{h} \rfloor}^{\lfloor \frac{k d_r}{h} \rfloor} I_{r,j}(L)(z')^{r+1+j} \right) (w')^{n-1-r}. \end{aligned} \tag{5.21}$$

Since $I_{n-1,-k'}(L) \neq 0$, the lowest order term in the above sum is given by

$$I_{n-1,-k'}(L)(z')^{n-k'}.$$

Hence it follows that P_∞ is a non-ordinary singular point of order $n-k'$.

(c) From [**vMM**], we know that $C(L)$ is irreducible for generic L (which corresponds to generic difference operator $\mathbf{C}$). To show that $C(L)$ has singular points only at P_0 and P_∞ generically, it suffices to show that the system of equations

$$I(L,z,w) = \frac{\partial I}{\partial z}(L,z,w) = \frac{\partial I}{\partial w}(L,z,w) = 0 \tag{5.22}$$

has only the solution $z=w=0$ for generic L. To show that this is the case, first observe that

(5.23)

$$I(L,0,w) = (-1)^n w^n, \ \frac{\partial I}{\partial z}(L,0,w) = I_{00}(L) w^{n-1}, \ \frac{\partial I}{\partial w}(L,0,w) = (-1)^n n w^{n-1}.$$

Therefore the system of equations $I(L,0,w) = \frac{\partial I}{\partial z}(L,0,w) = \frac{\partial I}{\partial w}(L,0,w) = 0$ has only the solution $w=0$ for any L. Next, we consider the system

$$I(L,z,0) = \frac{\partial I}{\partial z}(L,z,0) = \frac{\partial I}{\partial w}(L,z,0) = 0. \tag{5.24}$$

But since we have

$$\begin{aligned} I(L,z,0) &= \sum_{j=k'}^{k} I_{n-1,j} z^{n-j} = z^{n-k} \bar{I}_1(L,z), \ \frac{\partial I}{\partial z}(L,z,0) = z^{n-k-1} \bar{I}_2(L,z), \\ \frac{\partial I}{\partial w}(L,z,0) &= \sum_{j=-(k'-1)}^{k-1} I_{n-2,j} z^{n-1-j} = z^{n-k} \bar{I}_3(L,z). \end{aligned} \tag{5.25}$$

Therefore the system in (5.24) has $z=0$ as the only solution iff $\bar{I}_1(L,z) = \bar{I}_2(L,z) = \bar{I}_3(L,z) = 0$ has no solution. To get the condition for this, we use the method of Kronecker [**Kron, P, O**]. Accordingly, introduce the indeterminates u_1, u_2, u_3; v_1, v_2, v_3 and consider the resultant

$$\text{Res}\left(\sum_{i=1}^{3} u_i \bar{I}_i, \sum_{i=1}^{3} v_i \bar{I}_i\right) = \sum_{\alpha,\beta} u_1^{\alpha_1} u_2^{\alpha_2} u_3^{\alpha_3} v_1^{\beta_1} v_2^{\beta_2} v_3^{\beta_3} R_{\alpha,\beta}(L), \tag{5.26}$$

where $\alpha = (\alpha_1, \alpha_2, \alpha_3)$, $\beta = (\beta_1, \beta_2, \beta_3)$ are multi-indices. The fact is that the system $\bar{I}_1(L,z) = \bar{I}_2(L,z) = \bar{I}_3(L,z) = 0$ has a solution iff the above polynomial in u_i, v_j vanishes identically, which means $R_{\alpha,\beta}(L) = 0$ for all α, β. Therefore, if L is such that $R_{\alpha,\beta}(L) \neq 0$ for all α, β, then the system $\bar{I}_1(L,z) = \bar{I}_2(L,z) = \bar{I}_3(L,z) = 0$ will have no solution. Since $R_{\alpha,\beta}(L)$ are polynomials, this shows generically (5.24) has $z = 0$ as the only solution. Finally, to show that (5.22) has no solution in $(\mathbb{C}^*)^2$ generically, we will use the notion of A-discriminants, introduced in [**GZK**] (see also [**GKZ**] for an extensive discussion). For us, we take $A = \Delta(I) \cap \mathbb{N}^2$, the set of lattice points in the triangular region with vertices at $(0,n), (n-k,0)$, and $(n+k',0)$. Denote by $\mathbb{C}^A$ the finite dimensional space of polynomials in the form

$$f = \sum_{(a,b)\in A} f_{ab} z^a w^b, \quad f_{ab} \in \mathbb{C}. \tag{5.27}$$

Then the A-discriminant Δ_A is an irreducible polynomial in $\mathbb{Z}[(f_{ab})]$ having the property that if $f \in \mathbb{C}^A$ is such that there exists $(z_0, w_0) \in (\mathbb{C}^*)^2$ such that

$$f(z_0, w_0) = \frac{\partial f}{\partial z}(z_0, w_0) = \frac{\partial f}{\partial w}(z_0, w_0) = 0, \tag{5.28}$$

then $\Delta_A(f) = 0$. Consequently, if L is such that $\Delta_A(I(L,\cdot,\cdot)) \neq 0$, then it will follow that (5.22) has no solution in $(\mathbb{C}^*)^2$. This completes the proof. □

THEOREM 5.6. *Let $L \in \mathcal{G}$ satisfy the addtional genericity assumptions:*

(GA1) *$I(L,z,w)$ is irreducible,*

(GA2) *if $\bar{I}_1(L,z)$, $\bar{I}_2(L,z)$ and $\bar{I}_3(L,z)$ are defined by $I(L,z,0) = z^{n-k}\bar{I}_1(L,z)$, $\frac{\partial I}{\partial z}(L,z,0) = z^{n-k-1}\bar{I}(L,z)$, $\frac{\partial I}{\partial w}(L,z,0) = z^{n-k}\bar{I}_3(L,z)$, and $R_{\alpha,\beta}(L)$ are defined by (5.26), then $R_{\alpha,\beta}(L) \neq 0$ for all α, β,*

(GA3) *the resultant* $\text{Res}(\bar{I}_1(L,\cdot), \bar{I}_2(L,\cdot)) \neq 0$*, or equivalently, the map*

$$\pi : C(L) \longrightarrow \mathbb{CP}^1, [\xi : z : w] \mapsto [\xi : w] \tag{5.29}$$

has no ramification points of the form $[1 : z : 0]$.

(GA4) *$\Delta_A(I(L,\cdot,\cdot)) \neq 0$, where Δ_A is the A-resultant for A given by the set of lattice points in the triangular region with vertices at $(0,n)$, $(n-k,0)$ and $(n+k',0)$.*

Then we have the following results.

(a) *The genus of the nonsingular model $\widetilde{C}(L)$ of $C(L)$ is given by*

$$g(\widetilde{C}(L)) = \text{Card}(\Delta(I)^o \cap \mathbb{N}^2) = \frac{1}{2}[(n-1)(k+k') - d - d' + 2]. \tag{5.30}$$

(b) *If we denote the pullback of z and w by ϕ also by the same symbols, then a basis of holomorphic 1-forms of $\widetilde{C}(L)$ is given by*

$$\omega_{ab} = \frac{z^{a-1} w^{b-1}}{\partial I / \partial w} dz, \ (a,b) \in \Delta(I)^o \cap \mathbb{N}^2. \tag{5.31}$$

PROOF. In order to apply the Baker-Khovansii theorem, we have to establish the nondegeneracy of $I(L,z,w)$ with respect to the Newton polygon $\Delta(I)$. So let τ_1 be the edge joining $(0,n)$ and $(n-k,0)$, τ_2 the edge joining $(n-k,0)$ and $(n+k',0)$, and τ_3 the edge joining $(n+k',0)$ and $(0,n)$. Then

$$\begin{aligned} I_{\tau_1}(z,w) &= (-1)^n w^n + \sum_{j=1}^{d} I_{n_1 j-1,k_1 j}(L) z^{(n_1-k_1)j} w^{n_1(d-j)} \\ I_{\tau_2}(z,w) &= \sum_{j=-k'}^{k} I_{n-1,j}(L) z^{n-j}, \\ I_{\tau_3}(z,w) &= (-1)^n w^n + \sum_{j=1}^{d'} I_{n_1' j-1,-k_1' j}(L) z^{(n_1'+k_1')j} w^{n_1'(d'-j)}. \end{aligned} \tag{5.32}$$

As in [LN2], we have

$$I_{n_1 j-1,k_1 j}(L) = (-1)^{n-j} E_j(\sigma_1(L),\ldots,\sigma_d(L)), \tag{5.33}$$

where E_j is the j-th elementary symmetric function. Hence $I_{\tau_1}(L,z,w)$ is the weighted homogeneous polynomial

$$I_{\tau_1}(L,z,w) = (-1)^n \prod_{i=1}^{d} (w^{n_1} - \sigma_i(L) z^{n_1-k_1}). \tag{5.34}$$

From this formula, and the fact that L is on a generic coadjoint orbit, it is easy to see that $I_{\tau_1} = \frac{\partial I_{\tau_1}}{\partial z} = \frac{\partial I_{\tau_1}}{\partial w} = 0$ has no solution in $(\mathbb{C}^*)^2$. In a similar way, we have

$$I_{n_1' j-1,-k_1' j}(L) = (-1)^{n-j} E_j(\sigma_1'(L),\ldots,\sigma_{d'}'(L)), \tag{5.35}$$

and

$$I_{\tau_3}(L,z,w) = (-1)^n \prod_{i=1}^{d'} (w^{n_1'} - \sigma_i'(L) z^{n_1'+k_1'}). \tag{5.36}$$

So the system $I_{\tau_3} = \frac{\partial I_{\tau_3}}{\partial z} = \frac{\partial I_{\tau_3}}{\partial w} = 0$. has no solution in $(\mathbb{C}^*)^2$. On the other hand, the fact that $I_{\tau_2} = \frac{\partial I_{\tau_2}}{\partial z} = \frac{\partial I_{\tau_2}}{\partial w} = 0$ has no solution in $(\mathbb{C}^*)^2$ is a consequence of the assumption (GA3). Now if we take $\tau = \Delta(I)$, then the system $I_\tau = \frac{\partial I_\tau}{\partial z} = \frac{\partial I_\tau}{\partial w} = 0$ has no solution in $(\mathbb{C}^*)^2$ because $(0,0)$ is the only solution by (GA2) and (GA4) and it is not in $(\mathbb{C}^*)^2$. Finally, if we take τ to be the three vertices of the triangle $\partial\Delta(I)$, the nondegeneracy condition holds trivially.

To complete the proof, it remains to compute $\mathrm{Card}(\Delta(I)^o \cap \mathbb{N})$. To do so, we recall that if γ is the line segment between the integral points (a,b) and (c,d), then [BP]

$$i(\gamma) = \text{number of integral points on } \gamma = \gcd(a-c, b-d) + 1. \tag{5.37}$$

Thus it follows that

$$i(\gamma_1) = d+1,\ i(\gamma_2) = k+k'+1,\ i(\gamma_3) = d'+1 \tag{5.38}$$

and hence the number of lattice points on $\partial\Delta(I)$ is given by

$$i(\partial\Delta(I)) = k+k'+d+d'. \tag{5.39}$$

Finally, applying Pick's theorem, we find that

$$\begin{aligned}\operatorname{Card}(\Delta(I)^o \cap \mathbb{N}^2) &= \operatorname{Area}(\Delta(I)) - \frac{1}{2} i(\partial\Delta(I)) + 1 \\ &= \frac{1}{2}[(n-1)(k+k') - d - d' + 2].\end{aligned} \tag{5.40}$$

□

REMARK 5.7. Note that from Proposition 5.1 (b) and the proposition above, we have

$$\#\text{ of nontrivial integrals } I_{rj} - g(\widetilde{C}(L)) = d' - 1. \tag{5.41}$$

In the case of the periodic Kostant flow in [**LN2**], $k' = 1$ and so $d' - 1 = 0$. However, this number can be strictly positive. As the reader will see, in the case where $d' - 1 > 0$, there are (nontrivial) Hamiltonian flows generated by precisely $d' - 1$ coefficients of $I(L, z, w)$ which do not move on the Jacobi variety of the curve $\widetilde{C}(L)$. That is, these flows are not linearizable on the Jacobi variety. The problem of finding variables which move linearly under these flows is important not just from a dynamical systems point of view, but is also relevant in extending the dictionary in [**vMM**]. Geometrically, as the reader will see, we have the following picture. For the case $d' = 1$, generically, the level set of integrals embeds into a Jacobi variety, since the regular divisors form an open subset. However, if $d' > 1$, we need to make a $(\mathbb{C}^*)^{d'-1}$ extension of the Jacobi variety, corresponding to the d' points in the normalization which project to P_∞. In this connection, the reader is referred to Proposition 5.8, Theorem 5.12, (5.80), Corollary 5.14, Theorem 5.21, and Theorem 6.6 below for a quick glance of what to expect.

PROPOSITION 5.8.

(a) *The number of local branches of $C(L)$ at the singular point P_∞ is equal to d'.*
(b) *The number of local branches of $C(L)$ at the singular point P_0 is equal to d.*

PROOF.

(a) Let z', w' be the coordinates in the neighborhood of P_∞ introduced in the proof of Proposition 5.5 (b) and let

$$\widetilde{I}(L, z', w') = \det(L_1 + L_0 z' + L_{-1}(z')^2 - w' I). \tag{5.42}$$

Then the Newton polygon of $\widetilde{I}(L, z', w')$ is the triangular region with vertices at $(0, n)$, $(n - k', 0)$ and $(n + k, 0)$. Therefore, for the equation $\widetilde{I}(L, z', w') = 0$, the branches $z'(w')$ in a neighborhood of $(z', w') = (0, 0)$ are given by

$$z' = (w')^{\gamma_1}(c_1 + z'_1), \; z'_1 = c_2 (w')^{\gamma_2} + c_3 (w')^{\gamma_2 + \gamma_3} + \cdots \tag{5.43}$$

where $\gamma_1 = \frac{n}{n-k'}$ is the negative slope of the segment joining $(0, n)$ and $(n - k', 0)$. (See, for example, [**W**].) Substitute this in the equation $\widetilde{I}(L, z', w') = 0$ gives
(5.44)

$$(-1)^n (w')^n + \sum_{r=0}^{n-1} \left(\sum_{j=-\lfloor \frac{k' d_r}{h} \rfloor}^{\lfloor \frac{k d_r}{h} \rfloor} I_{r,j} (c_1 + z'_1)^{r+1+j} (w')^{\frac{k'(r+1)+nj}{n-k'}} \right) (w')^n = 0.$$

Hence the lowest order term in w in the above equation gives

$$
\begin{aligned}
0 &= (-1)^n (w')^n + \left(\sum_{j=1}^{d'} I_{n_1' j-1, -k_1' j}(L) c_1^{(n_1'-k_1')j} \right) (w')^n \\
&= (-1)^n \prod_{i=1}^{d'} (1 - \sigma_i'(L) c_1^{n_1'-k_1'})(w')^n
\end{aligned} \tag{5.45}
$$

and so $c_1^{n_1'-k_1'} = \frac{1}{\sigma_i'(L)}$, $i = 1, \ldots, d'$. As the σ_i''s are distinct, it follows that we have d' local branches of $C(L)$ at P_∞.

(b) The proof proceeds as in [**LN2**, Proposition 3.10 (b)]. □

We will henceforth denote by $\phi : \widetilde{C}(L) \longrightarrow C(L)$ the normalization of $C(L)$. Then in view of the above proposition, we have

$$
\begin{aligned}
\phi^{-1}(P_0) &= \{P_{0,1}, \ldots, P_{0,d}\}, \\
\phi^{-1}(P_\infty) &= \{P_{\infty,1}, \ldots, P_{\infty,d'}\}.
\end{aligned} \tag{5.46}
$$

The calculation in the proof of the above proposition also shows that

$$
\begin{aligned}
(z) &= n_1(P_{0,1} + \cdots + P_{0,d}) - n_1'(P_{\infty,1} + \cdots + P_{\infty,d'}), \\
(w) &= (n_1 - k_1)(P_{0,1} + \cdots + P_{0,d}) - (n_1' + k_1')(P_{\infty,1} + \cdots + P_{\infty,d'}) \\
&\quad + (\text{an effective divisor of degree } k + k').
\end{aligned} \tag{5.47}
$$

Following [**vMM**], we introduce the eigenvector $\bar{f} = (f_1, \ldots, f_{n-1}, z)^T$ of $zL(z)$ with last component equal to z so that $zL(z)\bar{f} = w\bar{f}$. We will extend f_i to all $i \in \mathbb{Z}$ with the condition that $f_{i+n} = zf_i$. Likewise, introduce the eigenvector $\bar{f}' = (f_1', \ldots, f_{n-1}', z^{-1})^T$ of $(zL(z))^T$ so that $(zL(z))^T\bar{f}' = w\bar{f}'$. We will extend f_i' to all $i \in \mathbb{Z}$ by $f_{i+n}' = z^{-1}f_i'$. Finally, we will define $P_{0,i}$ (resp. $P_{\infty,i}$) for all $i \in \mathbb{Z}$ by $P_{0,i+d} = P_{0,i}$ (resp. $P_{\infty,i+d'} = P_{\infty,i}$).

Then for $L \in \mathcal{G}$ (if $\mathbf{C}$ is the infinite periodic band matrix corresponding to L, then $\mathbf{C}$ is regular in the terminology in [**vMM**]), we have the following result [**vMM**, Proposition 1].

PROPOSITION 5.9. *Every $L \in \mathcal{G}$ leads to $2n$ regular*[1] *divisors $D^{(i)}(L)$ and $D_2^{(i)}(L)$ of degree $g(\widetilde{C}(L))$, $i = 1, \ldots, n$, such that*

$$
\begin{aligned}
(f_i) &= D^{(i)}(L) - D(L) - \sum_{a=1}^{i} P_{\infty,a} + \sum_{a=0}^{i-1} P_{0,a}, \quad D(L) = D^{(n)}(L) \\
(f_i') &= D_2^{(i)}(L) - D'(L) - \sum_{a=1}^{i} P_{0,a} + \sum_{a=0}^{i-1} P_{\infty,a}, \quad D'(L) = D_2^{(n)}(L).
\end{aligned} \tag{5.48}
$$

Indeed, the first step in showing $D(L)$ is regular consists of showing $D(L)$ is *general* in the sense that $\dim \mathscr{L}(D(L)) = 1$ [**vMM**] where

$$
\mathscr{L}(D(L)) = \{f \text{ meromorphic on } \widetilde{C}(L) \mid (f) \geq -D(L)\}.
$$

This fact will be used later on.

---

[1] A divisor D is called regular (with respect to $\{P_{0,i}\}_{i\in\mathbb{Z}}$ and $\{P_{\infty,j}\}_{j\in\mathbb{Z}}$) if $\dim \mathscr{L}\left(D + \sum_{i=1}^{j} P_{\infty,i} - \sum_{i=0}^{j} P_{0,i}\right) = 0$ for all j.

In terms of $\bar{f}$ and $\bar{f}'$, we define the normalized eigenvectors

$$\widehat{v}(P) = \frac{\bar{f}}{z}(P), \quad \widehat{v}'(P) = \frac{\bar{f}'}{\bar{f}'_1}(P). \tag{5.49}$$

Then from (5.48), we have

$$\begin{aligned} (\widehat{v}_i) &= D^{(i)}(L) - D(L) + \sum_{a=i+1}^{n} P_{\infty,a} - \sum_{a=i}^{n-1} P_{0,a}, \quad 1 \le i \le n-1. \\ (\widehat{v}'_i) &= D_2^{(i)}(L) - D'(L) + \sum_{a=1}^{i-1} P_{\infty,a} - \sum_{a=2}^{i} P_{0,a}, \quad 2 \le i \le n. \end{aligned} \tag{5.50}$$

For $L \in \mathcal{G}$ satisfying the genericity assumptions in Theorem 5.6, we now define

$$\phi_{ab}(L) = \int_{D_0}^{D(L)} \omega_{ab}, \quad (a,b) \in \Delta(I)^o \cap \mathbb{N}^2, \quad D_0 = gP_{0,1}, \tag{5.51}$$

where $g = g(\widetilde{C}(L))$. We will henceforth use this shorthand in our formulas. Clearly, these variables are defined modulo the lattice Λ in the space $\mathbb{C}^g$ generated by the columns of the $g \times 2g$ period matrix corresponding to the basis $\{\omega_{ab} \mid (a,b) \in \Delta(I)^o \cap \mathbb{N}^2\}$ of holomorphic 1-forms of $\widetilde{C}(L)$ and a canonical basis $\delta_1, \ldots, \delta_{2g}$ of $H_1(\widetilde{C}(L), \mathbb{Z})$. In other words, these variables are defined on the Jacobi variety of $\widetilde{C}(L)$.

In order to compute the Poisson bracket between the $I_{r,j}$ and the ϕ_{ab}, we begin with the following proposition.

PROPOSITION 5.10. *The Hamiltonian equations of motion generated by a smooth function $F(L)$ in the Lie-Poisson structure on $\mathfrak{s}^*$ are given by*

$$\begin{aligned} \dot{L}_0 &= \Pi_{\mathfrak{n}_-^\perp}\left(\left[\Pi_{\mathfrak{b}_+}\delta_0 F, L_0\right] + \left[\delta_{-1}F, L_{-1}\right]\right) \\ &\quad + \Pi_{\mathfrak{b}_+^\perp}\left(\left[L_0, \Pi_{\mathfrak{n}_-}\delta_0 F\right] + \left[L_1, \delta_1 F\right]\right), \\ \dot{L}_{-1} &= \left[\Pi_{\mathfrak{b}_+}\delta_0 F, L_{-1}\right], \\ \dot{L}_1 &= \left[L_1, \Pi_{\mathfrak{n}_-}\delta_0 F\right], \end{aligned} \tag{5.52}$$

where $\delta_0 F$, $\delta_{-1}F$ and $\delta_1 F$ are the derivatives of F with respect to L_0, L_{-1} and L_1 respectively, and where $\Pi_{\mathfrak{b}_+^\perp}$ and $\Pi_{\mathfrak{n}_-^\perp}$ are the projection operators relative to the decomposition $\mathfrak{g} = \mathfrak{b}_+^\perp \oplus \mathfrak{n}_-^\perp$, given by

$$\Pi_{\mathfrak{b}_+^\perp} A = A_+, \quad \Pi_{\mathfrak{n}_-^\perp} A = A_- + A_0. \tag{5.53}$$

The operators $\Pi_{\mathfrak{b}_+}$ and $\Pi_{\mathfrak{n}_-}$, on the other hand, are projection operators to $\mathfrak{b}_+$ and $\mathfrak{n}_-$, respectively, relative to $\mathfrak{g} = \mathfrak{n}_- \oplus \mathfrak{b}_+$.

PROOF. The result follows by a direct calculation from the expression for the Lie-Poisson structure on $\mathfrak{s}^*$:

$$\begin{aligned} &\{F_1, F_2\}_{\mathfrak{s}^*}(L_0, (L_{-1}, L_1)) \\ &= (L_0, [\,\delta_0 F_1, \delta_0 F_2\,]_J) + (L_{-1}, [\,\Pi_{\mathfrak{b}_+}\delta_0 F_1, \delta_{-1}F_2\,] + [\,\delta_{-1}F_1, \Pi_{\mathfrak{b}_+}\delta_0 F_2\,]) \\ &\quad + (L_1, [\,\Pi_{\mathfrak{n}_-}\delta_0 F_1, \delta_1 F_2\,] + [\,\delta_1 F_1, \Pi_{\mathfrak{n}_-}\delta_0 F_2\,]), \end{aligned} \tag{5.54}$$

where $\delta_0 F_i$, $\delta_{-1}F_i$ and $\delta_1 F_i$ are the derivatives of F_i with respect to L_0, L_{-1} and L_1 respectively, $i = 1, 2$. ☐

LEMMA 5.11. *Let* (z_0, w_0) *be such that* $[1 : z_0 : w_0] \notin C_{\mathrm{aff}}(L)$*, and let*

$$H_{z_0,w_0}(L_0, (L_{-1}, L_1)) = \log\det(L_1 z_0^2 + L_0 z_0 + L_{-1} - w_0 I). \tag{5.55}$$

Then under the Hamiltonian flow generated by H_{z_0,w_0}*, the loop* $L(z) = L_1 z + L_0 + L_{-1}z^{-1}$ *evolves according to the equation*

$$\dot{L}(z) = [\, B_{z_0,w_0}(z), L(z)\,], \quad z \neq z_0, \tag{5.56}$$

where

$$B_{z_0,w_0}(z) = \Pi_{\mathfrak{b}_+} z_0(z_0 L(z_0) - w_0 I)^{-1} - \frac{z z_0}{z - z_0}(z_0 L(z_0) - w_0 I)^{-1}. \tag{5.57}$$

PROOF. To simplify notation, let $A(z_0, w_0) = z_0(z_0 L(z_0) - w_0 I)^{-1}$. Then

$$\begin{aligned} \delta_1 H_{z_0,w_0} &= z_0 A(z_0, w_0), \quad \delta_0 H_{z_0,w_0} = A(z_0, w_0), \\ \delta_{-1} H_{z_0,w_0} &= z_0^{-1} A(z_0, w_0). \end{aligned} \tag{5.58}$$

Direct substitution in (5.52) then yields the Hamiltonian equation generated by H_{z_0,w_0}:

$$\begin{aligned} \dot{L}_0 &= \Pi_{\mathfrak{b}_-}\left([\,\Pi_{\mathfrak{b}_+} A(z_0, w_0), L_0\,] + [\,A(z_0, w_0), L_{-1} z_0^{-1}\,]\right) \\ &\qquad + \Pi_{\mathfrak{n}_+}\left([\,L_0, \Pi_{\mathfrak{n}_-} A(z_0, w_0)] + [\,L_1 z_0, A(z_0, w_0)\,]\right), \\ \dot{L}_{-1} &= [\,\Pi_{\mathfrak{b}_+} A(z_0, w_0), L_{-1}\,], \\ \dot{L}_1 &= [\,L_1, \Pi_{\mathfrak{n}_-} A(z_0, w_0)], \end{aligned} \tag{5.59}$$

where $\Pi_{\mathfrak{b}_-} = \Pi_{\mathfrak{n}_-^{\perp}}$ and $\Pi_{\mathfrak{n}_+} = \Pi_{\mathfrak{b}_+^{\perp}}$ are projection operators onto $\mathfrak{b}_-$ and $\mathfrak{n}_+$ respectively with respect to $\mathfrak{g} = \mathfrak{b}_- \oplus \mathfrak{n}_+$. Hence it follows from these equations that

$$\begin{aligned} \dot{L}(z) =&[\,\Pi_{\mathfrak{b}_+} A(z_0, w_0), L_{-1} z^{-1} + L_1 z\,] + [\,L_1 z, A(z_0, w_0)\,] \\ &+ [\,\Pi_{\mathfrak{b}_+} A(z_0, w_0), L_0\,] + [\,A(z_0, w_0), L_{-1} z_0^{-1}\,] \\ =&[\,\Pi_{\mathfrak{b}_+} A(z_0, w_0), L(z)\,] - [\,A(z_0, w_0), L_1(z + z_0) + L_0\,]. \end{aligned} \tag{5.60}$$

Next, it is easy to see that

$$L_1(z + z_0) + L_0 = \frac{z L(z) - z_0 L(z_0)}{z - z_0}, \quad z \neq z_0. \tag{5.61}$$

Substituting this into (5.60), we obtain the equation in (5.56), as asserted. □

Let

$$P_i = [1 : z_0 : w_i(L, z_0)] \in C_{\mathrm{aff}}(L) \setminus \{P_0\}, i = 1, \ldots, n. \tag{5.62}$$

In our next result, we will evaluate the Poisson brackets at a point $L \in \mathcal{G}$ satisfying the genericity assumptions in Theorem 5.6. For simplicity, we will denote the induced Poisson structure on $\mathcal{G} \subset \mathcal{T}^{(k,k')}$ again by $\{\cdot, \cdot\}$.

THEOREM 5.12. *For* $(a, b) \in \Delta(I)^o \cap \mathbb{N}^2$, $-\lfloor \frac{k' d_r}{h} \rfloor \leq j \leq \lfloor \frac{k d_r}{h} \rfloor$, $r = 1, \ldots, n - 1$*, we have*

$$\{\phi_{ab}, I_{r,j}\}(L) = \delta_{r,n-b}\delta_{j,n-a-b}. \tag{5.63}$$

PROOF. We will compute the Poisson bracket $\{\phi_{ab}, H_{z_0,w_0}\}(L)$, where L satisfies the genericity assumptions in Theorem 5.6. Indeed, in order to do so, it follows by continuity that it suffices to evaluate it on the (open) dense set of loops L satisfying the following additional conditions:

(i) the points of $D(L)$ are distinct, $\operatorname{supp} D(L) \subset C_{\mathrm{aff}}(L) \setminus \{P_0\}$,
(ii) $\operatorname{supp} D(L) \cap \{I_w(z,w) = 0\} = \emptyset$,
(iii) $\operatorname{supp} D(L) \cap \{P_i\}_{i=1}^n = \emptyset$,
(iv) $\{P_i\}_{i=1}^n \cap \{I_w(z,w) = 0\} = \emptyset$.

For small values of t, let $L(z,t)$ be the Hamiltonian flow generated by H_{z_0,w_0} with initial condition given by $L(z,t=0) = L(z)$ and let $D(t) = \sum_{j=1}^g Q_j(t)$ be the divisor of poles of the eigenvector of $zL(z,t) = L_1(t)z^2 + L_0(t)z + L_{-1}(t)$ with last component normalized to 1, $Q_j(t) = [1 : z_j(t) : w_j(t)]$, $Q_j = Q_j(0) = [1 : z_j : w_j]$, $j = 1, \ldots, g$. By our assumptions, for small values of t, the points $Q_j(t)$ are distinct. Moreover, we can use the coordinate function z as local parameter near each point Q_j. Hence we have

$$\{\phi_{ab}, H_{z_0,w_0}\}(L) = \sum_{j=1}^g \frac{z_j^{a-1} w_j^{b-1}}{I_w(z_j,w_j)} \left.\frac{dz_j(t)}{dt}\right|_{t=0}. \tag{5.64}$$

To compute $\left.\frac{dz_j(t)}{dt}\right|_{t=0}$, consider a representative $f(z,t)$ of f_t in a neighborhood of z_j for which $(e_n, f(z_j(t),t)) = 0$ for values of t close to 0. Differentiating this relation with respect to t at $t=0$, we find

$$\left.\frac{dz_j(t)}{dt}\right|_{t=0} = -\frac{\left(e_n, \frac{\partial f}{\partial t}(z_j,0)\right)}{\left(e_n, \frac{\partial f}{\partial z}(z_j,0)\right)}. \tag{5.65}$$

On the other hand, from differentiating

$$zL(z,t)f(z,t) = wf(z,t), \tag{5.66}$$

we obtain

$$(zL(z,t) - wI)\left(\frac{\partial f}{\partial t} - B_{z_0,w_0}(z,t)f(z,t)\right) = 0 \tag{5.67}$$

so that

$$\frac{\partial f}{\partial t}(z_j,0) = B_{z_0,w_0}(z_j)f(z_j) + \lambda_j f(z_j), \quad f(z_j) = f(z_j,0) \tag{5.68}$$

for some $\lambda_j \in \mathbb{C}$. As $(e_n, f(z_j)) = 0$, it follows from (5.68) that

$$\begin{aligned} &\left(e_n, \frac{\partial f}{\partial t}(z_j,0)\right) \\ &= \left(e_n, \left(\Pi_{\mathfrak{b}_+} z_0 (z_0 L(z_0) - w_0 I)^{-1} - \frac{z_j z_0}{z_j - z_0}(z_0 L(z_0) - w_0 I)^{-1}\right) f(z_j)\right) \\ &= -\frac{z_j z_0}{z_j - z_0}\left(e_n, (z_0 L(z_0) - w_0 I)^{-1} f(z_j)\right). \end{aligned} \tag{5.69}$$

When we insert (5.69) into (5.65), the result is

$$\left.\frac{dz_j(t)}{dt}\right|_{t=0} = \frac{z_j z_0}{z_j - z_0} \frac{(e_n, (z_0 L(z_0) - w_0 I)^{-1} f(z_j))}{(e_n, \frac{\partial f}{\partial z}(z_j,0))}. \tag{5.70}$$

Substituting this in (5.64), we obtain

$$\{\phi_{ab}, H_{z_0,w_0}\}(L) = \sum_{j=1}^g \frac{z_0 z_j^a w_j^{b-1}}{(z_j - z_0) I_w(z_j,w_j)} \frac{(e_n, (z_0 L(z_0) - w_0 I)^{-1} f(z_j))}{(e_n, \frac{\partial f}{\partial z}(z_j,0))}. \tag{5.71}$$

To evaluate this expression, consider the meromorphic 1-form

$$\frac{(e_n, \frac{zz_0}{z-z_0}(z_0L(z_0) - w_0I)^{-1}f(z))}{(e_n, f(z))}\omega_{ab} \tag{5.72}$$

which has poles at the points $P_i = [1 : z_0 : w_i(z_0)]$, $i = 1, \ldots, n$, $Q_j, j = 1, \ldots, g$, and possibly at $P_{\infty,1}, \ldots, P_{\infty,d'}$. Now

$$\begin{aligned} &\sum_{j=1}^{n} \mathrm{Res}_{P_i} \frac{(e_n, \frac{zz_0}{z-z_0}(z_0L(z_0) - w_0I)^{-1}f(z))}{(e_n, f(z))}\omega_{ab} \\ =&\sum_{j=1}^{n} \frac{z_0^{a+1}w_i(z_0)^{b-1}}{(w_i(z_0) - w_0)I_w(z_0, w_i(z_0))}. \end{aligned} \tag{5.73}$$

To compute the above sum, consider the 1-form $\frac{w^{b-1}}{(w-w_0)I(z_0,w)}\,dw$ which has simple poles at $w = w_0$ and $w = w_i(z_0)$, $i = 1, \ldots, n$. By computing the residues of this 1-form at its singularities, we find

$$\begin{aligned} &\sum_{j=1}^{n} \mathrm{Res}_{P_i} \frac{(e_n, \frac{zz_0}{z-z_0}(z_0L(z_0) - w_0I)^{-1}f(z))}{(e_n, f(z))}\omega_{ab} \\ =&z_0^{a+1}\left[\lim_{R\to\infty}\oint_{|w|=R}\frac{w^{b-1}}{(w-w_0)I(z_0,w)}\frac{dw}{2\pi\mathrm{i}} - \frac{w_0^{b-1}}{I(z_0,w_0)}\right] \\ =&-\frac{z_0^{a+1}w_0^{b-1}}{I(z_0, w_0)}. \end{aligned} \tag{5.74}$$

Therefore, by the residue theorem and (5.74), we obtain

$$\begin{aligned} &\{\phi_{ab}, H_{z_0,w_0}\}(L) \\ =&\sum_{j=1}^{g} \mathrm{Res}_{Q_j} \frac{(e_n, \frac{zz_0}{z-z_0}(z_0L(z_0) - w_0I)^{-1}f(z))}{(e_n, f(z))}\omega_{ab} \\ =&\frac{z_0^{a+1}w_0^{b-1}}{I(z_0, w_0)} - \sum_{j=1}^{d'} \mathrm{Res}_{P_{\infty,j}}\left(e_n, \frac{zz_0}{z-z_0}(z_0L(z_0) - w_0I)^{-1}\widehat{v}(z)\right)\omega_{ab}. \end{aligned} \tag{5.75}$$

To complete the proof, it remains to compute the second term on the last line of the above formula. From (5.50), the components of $\widehat{v}$ are holomorphic in some neighborhood $\mathcal{U}_j$ of $P_{\infty,j}$, $j = 1, \ldots, d'$. Indeed, it follows from (5.50) and the assumption $\mathrm{supp}\, D(L) \subset C_{\mathrm{aff}}(L) \setminus \{P_0\}$ that
(5.76)

$$\widehat{v}(P_{\infty,1}) \in \sum_{j=1}^{d'} \mathbb{C}e_{(n_1'-1)d'+i}, \ \widehat{v}(P_{\infty,2}) \in \sum_{i=2}^{d'} \mathbb{C}e_{(n_1'-1)d'+i}, \ldots, \widehat{v}(P_{\infty,d'}) = e_n.$$

In $\mathcal{U}_j$, introduce a local parameter t close to 0 such that $z = c_j' t^{-n_1'} + \cdots$, where $(c_j')^{n_1'+k_1'} = \frac{1}{\sigma_j'}$, then clearly, we have

$$\begin{aligned} &(e_n, \frac{zz_0}{z-z_0}(z_0L(z_0) - w_0I)^{-1}\widehat{v}(z))) \\ \sim&z_0(e_n, (z_0L(z_0) - w_0I)^{-1}\widehat{v}(P_{\infty,j})). \end{aligned} \tag{5.77}$$

This shows the meromorphic 1-form $(e_n, \frac{zz_0}{z-z_0}(z_0L(z_0) - w_0I)^{-1}\widehat{v}(z))\omega_{ab}$ has no singularity at $P_{\infty,j}$, $j = 1, \ldots, d'$. Consequently, it follows from (5.75) and the above argument that

$$\{\phi_{ab}, H_{z_0,w_0}\}(L) = \frac{z_0^{a+1}w_0^{b-1}}{I(z_0, w_0)}. \tag{5.78}$$

But on the other hand,

$$\begin{aligned}&\{\phi_{ab}, H_{z_0,w_0}\}(L)\\ =&\frac{1}{I(z_0, w_0)}\sum_{r=1}^{n-1}\sum_{j=-\lfloor\frac{k'd_r}{h}\rfloor}^{\lfloor\frac{kd_r}{h}\rfloor}\{\phi_{ab}, I_{r,j}\}z_0^{r+1-j}w_0^{n-r-1}.\end{aligned} \tag{5.79}$$

On comparing the last two expressions, the assertion follows. □

Note that for the case where $d' = 1$, the above result is sufficient to establish the integrability of (1.10) in the sense of Liouville on the generic coadjoint orbits in $\mathcal{G}$.

In the rest of the section, we will work with the case where $d' > 1$. In this case, it is immediate from the above theorem that for $(a, b) \in \Delta(I)^o \cap \mathbb{N}^2$, we have

$$\{\,\phi_{ab}, I_{n_1'\mu,-k_1'\mu}\,\}(L) = 0 \quad \text{for all } \mu = 1, \ldots, d' - 1. \tag{5.80}$$

Thus for $1 \le \mu \le d' - 1$, the Hamiltonian flow generated by $I_{n_1'\mu,-k_1'\mu}$ is not moving on the Jacobi variety of $\widetilde{C}(L)$ and therefore we have to introduce variables which move linearly under these flows. In what follows, instead of dealing with $I_{n_1'\mu,-k_1'\mu}$, we will find it more convenient to work with

$$J_i^\infty = \sum_{\mu=1}^{d'-1} I_{n_1'\mu,-k_1'\mu}(\sigma_i')^{d'-1-\mu}, \quad i = 1, \ldots, d' - 1. \tag{5.81}$$

The collection of variables $\{J_i^\infty\}_{i=1}^{d'-1}$ is equivalent to $\{I_{n_1'\mu,-k_1'\mu}\}_{\mu=1}^{d'-1}$ as we have

$$\begin{pmatrix} J_1^\infty \\ \vdots \\ J_{d'-1}^\infty \end{pmatrix} = V(\sigma_1', \ldots, \sigma_{d'-1}') \begin{pmatrix} I_{n_1'(d'-1),-k_1'(d'-1)} \\ \vdots \\ I_{n_1',-k_1'} \end{pmatrix}, \tag{5.82}$$

where $V(\sigma_1', \ldots, \sigma_{d'-1}') = ((\sigma_i')^{j-1})_{i,j=1}^{d'-1}$ is the Vandermonde matrix and the σ_i' are invariants of the coadjoint action.

To introduce variables which move linearly under the Hamiltonian flows generated the J_i^∞, we need to make use of the normalized eigenvectors $\widehat{v}(P)$ introduced in (5.49).

At this juncture, let us recall that if X is an $n \times n$ matrix, the Desnanot-Jacobi formula [**Br**] says that

$$\det X[\widehat{\alpha}|\widehat{\gamma}] \det X[\widehat{\beta}|\widehat{\delta}] - \det X[\widehat{\alpha}|\widehat{\delta}] \det X[\widehat{\beta}|\widehat{\gamma}] = \det X[\widehat{\alpha}, \widehat{\beta}|\widehat{\gamma}, \widehat{\delta}] \det X, \tag{5.83}$$

where $X[\widehat{\alpha}|\widehat{\gamma}]$ is the submatrix obtained from X by deleting its row α and column γ etc. Therefore, if $\Delta_{ij}(z, w)$ denotes the (i, j)-cofactor of the loop $zL(z) - wI$ with $P = [1 : z : w] \in C(L)$, it follows that

$$\widehat{v}_\lambda(P) = \frac{\Delta_{a\lambda}(P)}{\Delta_{an}(P)} \quad \text{for all } a, \quad \lambda = 1, \ldots, n, \tag{5.84}$$

and these relations extend to all $P \in \widetilde{C}(L)$ away from the poles.

In the rest of the section, setting $A = (a_{ij})_{i,j=1}^n := L(z=1)$, we will make the following additional genericity assumptions (without further mention):

(GA5) $\sum_{\mu=1}^{d'-1} \frac{\partial I_{n_1'\mu,-k_1'\mu}}{\partial a_{n-d'+i,n}} (\sigma_i')^{d-1-\mu} \neq 0$, $1 \le i \le d'-1$.

In our next result, we will study the asymptotic behaviour of various cofactors of the matrix loop $z(P)L(z(P)) - w(P)I$ as P approaches $P_{\infty,i}$. As before, in a neighborhood of $P_{\infty,i}$ on the curve, we parametrize points P using a parameter t such that $I(L, z(t), w(t)) = 0$, where

$$w(t) = t^{-(n_1'+k_1')}, \quad z(t) = c_i' t^{-n_1'} + \cdots, \quad \text{with } (c_i')^{n_1'+k_1'} = \frac{1}{\sigma_i'(L)}. \tag{5.85}$$

PROPOSITION 5.13.

(a) *For* $1 \le i \le d'-1$, $\lambda = n-d'+i+1, \ldots, n$, *we have*

(5.86)

$$\Delta_{n-d'+i,\lambda}(P) = \left(\sum_{\mu=1}^{d'-1} \frac{\partial I_{n_1'\mu,-k_1'\mu}}{\partial a_{n-d'+i,\lambda}} (\sigma_i')^{-\mu} \right) t^{-(n_1'+k_1')(n-1)} + O(t^{-(n_1'+k_1')(n-1)+1}),$$

as $P \to P_{\infty,i}$.

(b) *For* $1 \le i \le d'-1$,

$$\begin{aligned} &\Delta_{n-d'+i,n-d'+i}(P) \\ =&(-1)^{n-1} w^{n_1'-1} \prod_{a \neq i} (w^{n_1'} - \sigma_a' z^{n_1'+k_1'}) + \cdots \\ \sim&(-1)^{n-1} \prod_{a\neq i} \left(1 - \frac{\sigma_a'}{\sigma_i'}\right) t^{-(n_1'+k_1')(n-1)} \\ =&\left(\sum_{\mu=0}^{d'-1} \frac{\partial I_{n_1'\mu,-k_1'\mu}}{\partial a_{n-d'+i,n-d'+i}} (\sigma_i')^{-\mu} \right) t^{-(n_1'+k_1')(n-1)}, \quad \text{as } P \to P_{\infty,i}. \end{aligned} \tag{5.87}$$

(c) *For* $1 \le i \le d'-1$, $\lambda = n-d'+i+1, \ldots, n-1$, *we have*

$$\Delta_{\lambda n}(P) \sim \left(\sum_{\mu=1}^{d'-1} \frac{\partial I_{n_1'\mu,-k_1'\mu}}{\partial a_{\lambda n}} (\sigma_i')^{-\mu} \right) t^{-(n_1'+k_1')(n-1)}, \quad \text{as } P \to P_{\infty,i}. \tag{5.88}$$

(d) *For* $1 \le i \le d'-1$,

$$\begin{aligned} \Delta_{1n}(P) = &\sum_{\substack{(r,j) \\ k_1'r+n_1'j=1}} \frac{\partial I_{r,j}}{\partial a_{1n}} (c_i')^{r+1-j} t^{-(n_1'+k_1')(n-1)-(n_1'-1)} \\ &+ O(t^{-(n_1'+k_1')(n-1)-(n_1'-2)}), \quad \text{as } P \to P_{\infty,i}. \end{aligned} \tag{5.89}$$

PROOF.

(a) By differentiating the cofactor expansion of $\det(zL(z) - wI)$ along the $(n - d'+i)$-th row of $zL(z) - wI$ with respect to $a_{n-d'+i,\lambda}$, we have $\Delta_{n-d'+i,\lambda}(P) = \sum_{r,j} \frac{\partial I_{r,j}}{\partial a_{n-d'+i,\lambda}} z^{r-j} w^{n-1-r}$. Using the parametrization in t in a neighborhood of

$P_{\infty,i}$ above, it follows as $t \to 0$ that

$$\frac{\partial I_{r,j}}{\partial a_{n-d'+i,\lambda}} z(t)^{r-j} w(t)^{n-1-r} = \frac{\partial I_{r,j}}{\partial a_{n-d'+i,\lambda}} (c_i')^{r-j} t^{-(n_1'+k_1')(n-1)+k_1' r+n_1' j} + \text{ higher order terms in } t. \tag{5.90}$$

To establish the asymptotics, it suffices to show that for each r, $I_{r,j}$ is independent of $a_{n-d'+i,\lambda}$ for $k_1' r + n_1' j < 0$, i.e., for j satisfying $-\lfloor \frac{k' d_r}{h} \rfloor \le j < -\frac{k_1' r}{n_1'}$.

To do so, note that the expansion of $E_{r+1}(L(z))$ in z is a sum of terms of the form

$$(L_0)_{i_1 j_1} \cdots (L_0)_{i_a j_a} (L_1)_{\ell_1 m_1} \cdots (L_1)_{\ell_b m_b} z^b (L_{-1})_{p_1 q_1} \cdots (L_{-1})_{p_c q_c} z^{-c}, \tag{5.91}$$

where

$$\begin{aligned} & a+b+c = r+1, \\ & -k' \le i_1 - j_1, \ldots, i_a - j_a \le k, n-k' \le \ell_1 - m_1, \ldots, \ell_b - m_b \le n-1, \\ & n-k \le q_1 - p_1, \ldots, q_c - p_c \le n-1, \end{aligned} \tag{5.92}$$

and moreover,

$$(j_1 - i_1) + \cdots (j_a - i_a) + (m_1 - \ell_1) + \cdots + (m_b - \ell_b) + (q_1 - p_1) + \cdots + (q_c - p_c) = 0. \tag{5.93}$$

To get the terms for $I_{r,j}(L)$, we restrict to those b and c such that $j = c - b$. Then the relation $a+b+c = r+1$ gives $j = r+1-a-2b$. Therefore

$$\begin{aligned} j < -\frac{k_1'}{n_1'} r & \iff n_1'(r+1-a-2b) \le -1 - k_1' r \\ & \iff n(r+1-a-2b) \le -d' - k' r \\ & \iff k' r + d' \le n(r+1-a-2c). \end{aligned} \tag{5.94}$$

On the other hand, (5.93) gives

$$\begin{aligned} & (\ell_1 - m_1) + \cdots + (\ell_b - m_b) - (q_1 - p_1) - \cdots - (q_c - p_c) \\ = & (j_1 - i_1) + \cdots + (j_a - i_1). \end{aligned} \tag{5.95}$$

Now from (5.92), we have the obvious estimate

$$b(n-k') - c(n-1) \le (\ell_1 - m_1) + \cdots + (\ell_b - m_b) - (q_1 - p_1) - \cdots - (q_c - p_c). \tag{5.96}$$

But on the other hand, by using (5.92) and then (5.94), we find

$$\begin{aligned} & b(n-k') - c(n-1) \\ & = (r+1-a-c)(n-k') - c(n-1) \\ & = (r+1-a-2c)(n-k') - c(k'-1) \\ & \ge k' r + d' - k'(r+1-a-2c) - c(k'-1) \\ & = k'(a-1) + d' + c(k'+1) \\ & \ge k'(a-1) + d'. \end{aligned} \tag{5.97}$$

Therefore, when we combine (5.96) and (5.97) and on using (5.95), we find

$$k'(a-1) + d' \le (j_1 - i_1) + \cdots + (j_a - i_a). \tag{5.98}$$

Now, if $a_{n-d'+i,\lambda}$ appeared in a term of the form in (5.91) with $c-b=j$, we would have

$$\begin{aligned} &\max[(j_1-i_1)+\cdots+(j_a-i_a)] \\ &\leq (\lambda-(n-d'+i))+k'(a-1) \\ &\leq d'-1+k'(a-1), \end{aligned} \tag{5.99}$$

which is clearly a contradiction to the inequality in (5.98). This contradiction shows that $I_{r,j}$ must be independent of $a_{n-d'+i,\lambda}$ for (r,j) satisfying $-\lfloor \frac{k'd_r}{h} \rfloor \leq j < -\frac{k_1 r}{n_1'}$.
(b) As in (a), we have $\Delta_{n-d'+i,n-d'+i}(P) = \sum_{r,j} \frac{\partial I_{r,j}}{\partial a_{n-d'+i,n-d'+i}} z^{r-j} w^{n-1-r}$. We want to show that for each r, $\frac{\partial I_{r,j}}{\partial a_{n-d'+i,n-d'+i}} = 0$ for those j for which $n_1'j+k_1'r<0$. Since this is the same condition as in (a), it follows that the inequality in (5.98) is valid. Now if $a_{n-d'+i,n-d'+i}$ appeared in (5.91) with $c-b=j$, we would have

$$(j_1-i_1)+\cdots+(j_a-i_a) \leq k'(a-1), \tag{5.100}$$

which is clearly a contradiction to the inequality in (5.98). Therefore if we express $\Delta_{n-d'+i,n-d'+i}(z,w)$ as a sum of quasi-homogeneous polynomials with weights n_1' and $n_1'+k_1'$, the highest degree term is given by

$$\sum_{\mu=0}^{d'-1} \frac{\partial I_{n_1'\mu,-k_1'\mu}}{\partial a_{n-d'+i,n-d'+i}} z^{(n_1'+k_1')\mu} w^{n-1-n_1'\mu}. \tag{5.101}$$

But by direct calculation, the highest degree term in $\Delta_{n-d'+i,n-d'+i}(z,w)$ is

$$(-1)^{n-1} w^{n_1'-1} \prod_{a\neq i} (w^{n_1'} - \sigma_a' z^{n_1'+k_1'}). \tag{5.102}$$

On equating the two expressions, we obtain the relation

$$\sum_{\mu=0}^{d'-1} \frac{\partial I_{n_1'\mu,-k_1'\mu}}{\partial a_{n-d'+i,n-d'+i}} z^{(n_1'+k_1')\mu} w^{n-1-n_1'\mu} = (-1)^{n-1} w^{n_1'-1} \prod_{a\neq i} (w^{n_1'} - \sigma_a' z^{n_1'+k_1'}) \tag{5.103}$$

from which the asymptotics follow.
(c) Analogous to part (a), for each r, we need to show $\frac{\partial I_{r,j}}{\partial a_{\lambda n}} = 0$ for those j for which $n_1'j+k_1'r<0$. Since $I_{r,j}(L)$ is a sum of terms of the form in (5.91) with $c-b=j$, the inequality in (5.98) is again valid. If $a_{\lambda n}$ appeared in (5.91) with $c-b=j$, we would have

$$\begin{aligned} &\max[(j_1-i_1)+\cdots+(j_a-i_a) \\ &\leq (n-\lambda)+k'(a-1) \\ &< (n-(n-d'+1))+k'(a-1) \\ &= (d'-1)+k'(a-1). \end{aligned} \tag{5.104}$$

This is clearly a contradiction to the inequality in (5.98)
(d) We have

$$\Delta_{1n}(P) = \sum_{r,j} \frac{\partial I_{r,j}}{\partial a_{1n}} z^{r+1-j} w^{n-1-r}. \tag{5.105}$$

As $t \to 0$,

$$\frac{\partial I_{r,j}}{\partial a_{1n}} z(t)^{r+1-j} w(t)^{n-1-r} = \frac{\partial I_{rj}}{\partial a_{1n}} (c_i')^{r+1-j} t^{-(n_1'+k_1')(n-1)+(k_1' r+n_1' j)-n_1'} + O(t^{-(n_1'+k_1')(n-1)+(k_1' r+n_1' j)-n_1'+1}) \tag{5.106}$$

We will show that for each r, $I_{r,j}$ is independent of a_{1n} for $-\lfloor \frac{k_1' d_r}{n_1'} \rfloor \leq j \leq -\frac{k_1'}{n_1'} r$, i.e., the partial derivative $\frac{\partial I_{r,j}}{\partial a_{1n}} = 0$ when $n_1' j + k_1' r \leq 0$.

Again we consider terms of the form in (5.91) satisfying (5.92) and (5.93) with $c - b = j$. This time, if a_{1n} appears in (5.91) with $j = c - b$, we would have

$$\max[(q_1 - p_1) + \cdots + (q_c - p_c)] \leq (n-1) + (c-1)(n-1) = (n-1)c. \tag{5.107}$$

On the other hand, (5.93) gives

$$\begin{aligned} &(q_1 - p_1) + \cdots + (q_c - p_c) \\ =&(\ell_1 - m_1) + \cdots + (\ell_b - m_b) + (i_1 - j_1) + \cdots + (i_a - j_a). \end{aligned} \tag{5.108}$$

Now from (5.92), we have the estimate

$$b(n-k') - ak' \leq (\ell_1 - m_1) + \cdots + (\ell_b - m_b) + (i_1 - j_1) + \cdots + (i_a - j_a). \tag{5.109}$$

But by the first line of (5.92) and the fact that $j \leq -\frac{k_1'}{n_1'} r \iff k' r \leq n(r+1-a-2c)$ (cf. (5.94)), we have

$$\begin{aligned} &b(n-k') - ak' \\ &= (r+1-a-c)(n-k') - ak' \\ &= n(r+1-a-c) - k'(r+1-c) \\ &= n(r+1-a-2c) + nc - k'(r+1-c) \\ &\geq k'r + nc - k'(r+1-c) \\ &= (n-1)c + c + k'(c-1) \\ &> (n-1)c. \end{aligned} \tag{5.110}$$

When we combine this with (5.109), it follows by (5.108) that

$$(n-1)c < (q_1 - p_1) + \cdots + (q_c - p_c), \tag{5.111}$$

which is a contraction to the inequality in (5.107). This contraction shows that we must have (5.89). □

COROLLARY 5.14. *For* $1 \leq i \leq d' - 1$,

$$\widehat{v}_{n-d'+i}(P_{\infty,i}) = \frac{\Delta_{n-d'+i,n-d'+i}}{\Delta_{n-d'+i,n}}(P_{\infty,i}) = \frac{(-1)^{n-1} \prod_{a \neq i}(\sigma_i' - \sigma_a')}{\sum_{\mu=1}^{d'-1} \frac{\partial I_{n_1'\mu,-k_1'\mu}}{\partial a_{n-d'+i,n}} (\sigma_i')^{d'-1-\mu}}, \tag{5.112}$$

$$\widehat{v}_\lambda(P_{\infty,i}) = \frac{\Delta_{n-d'+i,\lambda}}{\Delta_{n-d'+i,n}}(P_{\infty,i}) = \frac{\sum_{\mu=1}^{d'-1} \frac{\partial I_{n_1'\mu,-k_1'\mu}}{\partial a_{n-d'+i,\lambda}} (\sigma_i')^{d'-1-\mu}}{\sum_{\mu=1}^{d'-1} \frac{\partial I_{n_1'\mu,-k_1'\mu}}{\partial a_{n-d'+i,n}} (\sigma_i')^{d'-1-\mu}}, \quad \lambda \geq n - d' + i + 1.$$

Thus

$$\widehat{v}(P_{\infty,i}) = \sum_{\lambda=n-d'+i}^{n-1} \widehat{v}_\lambda(P_{\infty,i})e_\lambda + e_n, \quad 1 \le i \le d'-1 \tag{5.113}$$

and the condition for their existence is equivalent to the condition that $\operatorname{supp} D(L) \cap \{P_{\infty,1}, \ldots, P_{\infty,d'}\} = \emptyset$.

PROOF. Here we pick $a = n-d'+i$ in (5.84) and use the asymptotics in (a) and (b) of the above proposition to obtain the formulas for $\widehat{v}_\lambda(P_{\infty,i})$, $n-d'+i \le \lambda \le n$. By using the first line of (5.50), the assertion on $\widehat{v}(P_{\infty,i})$ is now clear. □

REMARK 5.15.

(a) The special choice of a in (5.84) in the evaluation of $\widehat{v}(P_{\infty,i})$ in the above corollary is critical. We remark that other choices have led us to nowhere.
(b) Clearly $\widehat{v}_{n-d'+i}(P_{\infty,i}) \neq 0$, $i = 1, \ldots, d'-1$. The logarithms of these quantities are the additional set of variables and we will eventually compute the Poisson brackets of these variables with the J_i^∞.

PROPOSITION 5.16. *The decomposition of* $\Delta_{nn}(z,w)$ *as a sum of quasi-homogeneous polynomials of weights* n_1' *and* $n_1'+k_1'$ *in decreasing degree is given by*

$$\begin{aligned}\Delta_{nn}&(z,w)\\ =&(-1)^{n-1}w^{n_1'-1}\prod_{a\neq d'}(w^{n_1'} - \sigma_a'(L)z^{n_1'+k_1'}) + \sum_{\substack{(r,j)\\ k_1'r+n_1'j=1}} \frac{\partial I_{r,j}}{\partial a_{nn}} z^{r-j}w^{n-1-r} + \cdots,\end{aligned} \tag{5.114}$$

where the first term is of degree $(n_1'+k_1')(n-1)$ *and the second term is of degree* $(n_1'+k_1')(n-1)-1$.

PROOF. The details are similar to the proof of Proposition 5.13 (b), (d) and so we will leave it to the reader. □

In order to formulate the next proposition, we have to introduce further notation. For $1 \le j \le d'$, let $Q_{j,d'}$ denote the set of sequences $\beta = (\beta_1, \ldots, \beta_j)$ of j integers for which $1 \le \beta_1 < \cdots < \beta_j \le d'$. If $\beta \in Q_{j,d'}$, we put

$$\Gamma_\beta' = \bigcup_{a=1}^{j} \{\beta_a, \beta_a + d', \ldots, \beta_a + (n_1'-1)d'\}. \tag{5.115}$$

PROPOSITION 5.17. *For* $\mu = 1, \ldots, d'-1$,

$$I_{n_1'\mu,-k_1'\mu}(L) = (-1)^{\mu-1} \sum_{\beta\in Q_{\mu,d'}} \prod_{a=1}^{\mu} \sigma_{\beta_a}' \left(\sum_{i\notin\Gamma_\beta'} a_{ii}\right) + \text{“other terms”}, \tag{5.116}$$

where “other terms” do not depend on $a_{11}, \ldots, a_{nn}$.

PROOF. The expansion of $E_{n_1'\mu+1}(L(z))$ in z consists of a sum of terms of the form

$$(L_0)_{i_1j_1}\cdots(L_0)_{i_aj_a}(L_1)_{\ell_1m_1}\cdots(L_1)_{\ell_bm_b}z^b(L_{-1})_{p_1q_1}\cdots(L_{-1})_{p_cq_c}z^{-c}, \tag{5.117}$$

where

$$
\begin{aligned}
&a+b+c = n_1'\mu + 1,\\
&-k' \le i_1 - j_1, \ldots, i_a - j_a \le k, n-k' \le \ell_1 - m_1, \ldots, \ell_b - m_b \le n-1, \\
&n-k \le q_1 - p_1, \ldots, q_c - p_c \le n-1,
\end{aligned}
\tag{5.118}
$$

and moreover,

$$
(j_1-i_1)+\cdots(j_a-i_a)+(m_1-\ell_1)+\cdots+(m_b-\ell_b)+(q_1-p_1)+\cdots+(q_c-p_c)=0. \tag{5.119}
$$

The coefficient $I_{n_1'\mu,-k_1'\mu}$ therefore consists of a sum of terms of the above form with $b-c=k_1'\mu$. For $1 \le i \le d'$, introduce

$$
\begin{aligned}
\widetilde{\sigma}_i'(L) &= \prod_{j=1}^{n_1'-k_1'} (L(z))_{i+(j-1)d',k'+i+(j-1)d'} \prod_{j=1}^{k_1'} (L(z))_{n-k'+i+(j-1)d',i+(j-1)d'} \\
&= \sigma_i'(L) z^{k_1'}.
\end{aligned}
\tag{5.120}
$$

If for an $\alpha \in Q'_{\mu,d}$, we multiply $\widetilde{\sigma}'_{\alpha_1}, \ldots, \widetilde{\sigma}'_{\alpha_\mu}$ together, then multiply by an element on the diagonal of L_0, and multiply by the corresponding signature of the permutation, we would obtain a term in $I_{n_1'\mu,-k_1'\mu}$. Note that for the product which we formed above, we have $a=(n_1'-k_1')\mu+1$, $b=k_1'\mu$ and $c=0$ so that $a+b+c=n_1'\mu+1$. The reason why we know that we need to multiply $\prod_{a=1}^{\mu}\sigma'_{\alpha_a}$ by an entry on the diagonal of L_0 is due the condition in (5.118), as can be easily verified. Now since $n_1'\mu$ rows and columns of $L(z)$ have been taken in forming the product $\prod_{a=1}^{\mu}\widetilde{\sigma}_{\alpha_a}(L)$, there are $n-n_1'\mu$ choices for the diagonal entry of L_0. This argument shows that $I_{n_1'\mu,-k_1'\mu}(L)$ is of the form in (5.116) and it remains to show that the "other terms" do not depend on the diagonal entries of L_0. To do so, note that as in the proof of Proposition 5.13 (a), we have the inequality

$$
b(n-k') - c(n-1) \le (j_1 - i_1) + \cdots + (j_a - i_a). \tag{5.121}
$$

On the other hand, it follows from $b-c=k_1'\mu$ and $a+b+c=n_1'\mu+1$ that

$$
\begin{aligned}
b(n-k') - c(n-1) &= n(b-c) - bk' + c \\
&= nk_1'\mu - bk' + c \\
&= k'(a+c-1) + c \\
&= k'(a-1) + (k'+1)c.
\end{aligned}
\tag{5.122}
$$

Combining the inequality in (5.121) with the relation in (5.122), we find

$$
k'(a-1) + (k'+1)c \le (j_1-i_1) + \cdots + (j_a - i_a). \tag{5.123}
$$

If we have x distinct diagonal entries of L_0 in the expansion (5.117), we would have the estimate

$$
\max[(j_1-i_1)+\cdots+(j_a-i_a)] \le k'(a-x). \tag{5.124}
$$

Therefore if $x \ge 1$, this will contradict the inequality in (5.123) as $c > 0$ for the "other terms". This contradiction establishes the assertion. □

By differentiating the formulas in the above proposition by $a_{\lambda\lambda}$, we obtain the following corollary.

COROLLARY 5.18. *For $1 \le \mu \le d'-1$, $1 \le \lambda \le n$, $\frac{\partial I_{n_1'\mu,-k_1'\mu}}{\partial a_{\lambda\lambda}}$ is given by*

$$\frac{\partial I_{n_1'\mu,-k_1'\mu}}{\partial a_{\lambda\lambda}} = (-1)^{\mu-1} \sum_{\substack{\beta\in Q_{\mu,d'}\\ \lambda\notin\Gamma'_\beta}} \prod_{a=1}^{\mu} \sigma'_{\beta_a}. \tag{5.125}$$

In particular, if $n-d'+i \le \lambda \le n$, $1 \le i \le d'-1$, we have

$$\frac{\partial I_{n_1'\mu,-k_1'\mu}}{\partial a_{\lambda\lambda}} = (-1)^{\mu-1} \sum_{\substack{\beta\in Q_{\mu,d'}\\ \lambda\notin\{\beta_1+(n_1'-1)d',\dots,\beta_\mu+(n_1'-1)d'\}}} \prod_{a=1}^{\mu} \sigma'_{\beta_a}. \tag{5.126}$$

PROPOSITION 5.19. *For $\lambda = n-d'+i+1, \dots, n$, $i = 1, \dots, d'-1$, we have*

$$\sum_{\mu=0}^{d'-1} \frac{\partial I_{n_1'\mu,-k_1'\mu}}{\partial a_{\lambda\lambda}} (\sigma_i')^{-\mu} = 0. \tag{5.127}$$

PROOF. By assumption, $\lambda > n-d'+i$. By Proposition 5.13 (b), we have

$$\begin{aligned}\Delta_{\lambda\lambda}(P) =& (-1)^{n-1} w^{n_1'-1} \prod_{a\neq\lambda} (w^{n_1'} - \sigma_a' z^{n_1'+k_1'}) + \text{terms of lower degree}\\ =& \sum_{r,j} \frac{\partial I_{r,j}}{\partial a_{\lambda\lambda}} z^{r-j} w^{n-1-r},\end{aligned} \tag{5.128}$$

where $\frac{\partial I_{r,j}}{\partial a_{\lambda\lambda}} = 0$ for (r,j) satisfying $-\lfloor \frac{k'd_r}{h} \rfloor \le j < -\frac{k'r}{h}$. In a neighborhood of $P_{\infty,i}$, if we make the substitution (5.85) into the above expression, the coefficient of $t^{-(n_1'+k_1')(n-1)}$ from the first expression is equal to 0, while the coefficient of $t^{-(n_1'+k_1')(n-1)}$ from the second expression is $\sum_{\mu=0}^{d-1} \frac{\partial I_{n_1'\mu,-k_1'\mu}}{\partial a_{\lambda\lambda}} \sigma_i^{-\mu}$. Hence the assertion follows. □

PROPOSITION 5.20. *For $1 \le \mu \le d'-1$, $1 \le i \le d'-1$, we have*

$$\{\log \widehat{v}_{n-d'+i}(P_{\infty,i}), I_{n_1'\mu,-k_1'\mu}\}(L) = \sum_{\lambda=n-d'+i}^{n-1} \widehat{v}_\lambda(P_{\infty,i}) \frac{\partial I_{n_1'\mu,-k_1'\mu}}{\partial a_{\lambda n}}. \tag{5.129}$$

PROOF. We will compute the Poisson bracket $\{\log \widehat{v}_{n-d'+i}(P_{\infty,i}), H_{z_0,w_0}\}(L)$. As in the proof of Theorem 5.12, for small values of t, let $L(z,t)$ be the solution of the Hamiltonian flow generated by H_{z_0,w_0} with initial condition given by $L(z,t=0) = L(z)$. Let $\bar{f}(P,t)$ be the eigenvector of $L(z,t)$ with last component equal to z. With $\widehat{v}(P,t) = \frac{\bar{f}(P,t)}{z}$, we have

$$\left.\frac{d}{dt}\right|_{t=0} \widehat{v}(P,t) = B_{z_0,w_0}(z(P))\widehat{v}(P) + \lambda(P)\widehat{v}(P) \tag{5.130}$$

for some scalar function $\lambda(P)$, where $B_{z_0,w_0}(z)$ is the loop given in (5.57). Since the last component of $\widehat{v}(P,t)$ is 1, it follows from the above equation that

$$\begin{aligned}0 =& (e_n, B_{z_0,w_0}(z)\widehat{v}(P)) + \lambda(P)\\ =& \lambda(P) + ((\Pi_{\mathfrak{b}_+} A(z_0,w_0))^T e_n, \widehat{v}(P)) - \frac{z(P)}{z(P)-z_0}(A(z_0,w_0)^T e_n, \widehat{v}(P)).\end{aligned} \tag{5.131}$$

Consequently,

$$\lambda(P) = -(e_n, A(z_0,w_0)e_n) + \frac{z(P)}{z(P)-z_0}(A(z_0,w_0)^T e_n, \widehat{v}(P)). \tag{5.132}$$

Since

$$A(z_0,w_0)^T e_n = \frac{z_0}{I(z_0,w_0)} \sum_{\lambda=1}^{n} \Delta_{\lambda n}(z_0,w_0) e_\lambda, \tag{5.133}$$

on letting $P \to P_{\infty,i}$ in the above expression, we find

$$\begin{aligned} \lambda(P_{\infty,i}) =& -(e_n, A(z_0,w_0)e_n) + (A(z_0,w_0)^T e_n, \widehat{v}(P_{\infty,i})) \\ =&(A(z_0,w_0)^T e_n, \sum_{\lambda=n-d'+i}^{n-1} \widehat{v}_\lambda(P_{\infty,i}) e_\lambda) \\ =&\frac{z_0}{I(z_0,w_0)} \sum_{\lambda=n-d'+i}^{n-1} \widehat{v}_\lambda(P_{\infty,i}) \Delta_{\lambda n}(z_0,w_0). \end{aligned} \tag{5.134}$$

On the other hand, by letting $P \to P_{\infty,i}$ in (5.130), the result is

$$\left.\frac{d}{dt}\right|_{t=0} \widehat{v}(P_{\infty,i},t) = -(\Pi_{\mathfrak{n}_-} A(z_0,w_0))\widehat{v}(P_{\infty,i}) + \lambda(P_{\infty,i})\widehat{v}(P_{\infty,i}). \tag{5.135}$$

Therefore,

$$\begin{aligned} &\{\log \widehat{v}_{n-d'+i}(P_{\infty,i}), H_{z_0,w_0}\}(L) \\ =&\frac{1}{\widehat{v}_{n-d'+i}(P_{\infty,i})} \left.\frac{d}{dt}\right|_{t=0} \widehat{v}_{n-d'+i}(P_{\infty,i},t) \\ =&\lambda(P_{\infty,i}) - \frac{1}{\widehat{v}_{n-d'+i}(P_{\infty,i})}((\Pi_{\mathfrak{n}_-} A(z_0,w_0))^T e_{n-d'+i}, \widehat{v}(P_{\infty,i})) \\ =&\lambda(P_{\infty,i}). \end{aligned} \tag{5.136}$$

On substituting (5.134) into (5.136), we obtain
(5.137)

$$\{\log \widehat{v}_{n-d'+i}(P_{\infty,i}), H_{z_0,w_0}\}(L) = \frac{z_0}{I(z_0,w_0)} \sum_{\lambda=n-d'+i}^{n-1} \widehat{v}_\lambda(P_{\infty,i}) \Delta_{\lambda n}(z_0,w_0).$$

The assertion then follows by comparing the coefficients of $\frac{z_0^{(n_1'+k_1')\mu+1} w_0^{n-1-n_1'\mu}}{I(z_0,w_0)}$ on both sides in (5.137) □

THEOREM 5.21. *For $1 \le i \le d'-1$, $1 \le j \le d'-1$, we have*

$$\begin{aligned} \{\log \widehat{v}_{n-d'+i}(P_{\infty,i}), J_j^\infty\}(L) =&\frac{\partial J_i^\infty}{\partial a_{n-d'+i,n-d'+i}} \delta_{ij} \\ =&(-1)^{n-1} \prod_{\substack{a=1 \\ a\neq i}}^{d'} (\sigma_i'(L) - \sigma_a'(L)) \delta_{ij}. \end{aligned} \tag{5.138}$$

PROOF. From (5.129), we obtain

$$\{\log \widehat{v}_{n-d'+i}(P_{\infty,i}), J_j^\infty\}(L) = \sum_{\lambda=n-d'+i}^{n-1} \widehat{v}_\lambda(P_{\infty,i}) \frac{\partial J_j^\infty}{\partial a_{\lambda n}}. \tag{5.139}$$

We divide the proof into three cases.

CASE 1. $i = j$

Using Corollary 5.14 in (5.139), we have

$$\{\log \widehat{v}(P_{\infty,i}), J_i^{\infty}\}(L) = \sum_{\lambda=n-d'+i}^{n-1} \frac{\Delta_{n-d'+i,\lambda}}{\Delta_{n-d'+i,n}}(P_{\infty,i}) \frac{\partial J_i^{\infty}}{\partial a_{\lambda n}}. \tag{5.140}$$

On the other hand, by applying the Desnanot-Jacobi formula, we find

$$\frac{\Delta_{n-d'+i,\lambda}}{\Delta_{n-d'+i,n}}(P)\Delta_{\lambda n}(P)(\sigma_i')^{d'-1} = \Delta_{\lambda\lambda}(P)(\sigma_i')^{d'-1}, \quad P \in \widetilde{C}(L), \tag{5.141}$$

for $\lambda = n - d' + i + 1, \ldots, n - 1$. For P in a neighborhood of $P_{\infty,i}$, we introduce a parameter t so that $P = P(t) = (z(t), w(t))$, where $z(t)$, $w(t)$ are given by the expressions in (5.85). Inserting this in (5.141), expand in t, it follows from the asymptotics as $t \to 0$ in Proposition 5.13 that

$$\widehat{v}_{\lambda}(P_{\infty,i}) \frac{\partial J_i^{\infty}}{\partial a_{\lambda n}} = \frac{\partial J_i^{\infty}}{\partial a_{\lambda\lambda}}. \tag{5.142}$$

Next we sum over both sides of the above expression from $\lambda = n-d'+i$ to $\lambda = n-1$, this gives

$$\sum_{\lambda=n-d'+i}^{n-1} \widehat{v}_{\lambda}(P_{\infty,i}) \frac{\partial J_i^{\infty}}{\partial a_{\lambda n}} = \sum_{\lambda=n-d'+i}^{n-1} \frac{\partial J_i^{\infty}}{\partial a_{\lambda\lambda}} = \frac{\partial J_i^{\infty}}{\partial a_{n-d'+i,n-d'+i}}, \tag{5.143}$$

where we have used Proposition 5.17 and Proposition 5.19. Substituting into (5.140) and using Proposition 5.13 (b) , we obtain

$$\{\log \widehat{v}_{n-d'+i}(P_{\infty,i}), J_i^{\infty}\}(L) = \frac{\partial J_i^{\infty}}{\partial a_{n-d'+i,n-d'+i}} = (-1)^{n-1} \prod_{\substack{a=1 \\ a \neq i}}^{d'} (\sigma_i'(L) - \sigma_a'(L)), \tag{5.144}$$

as asserted.

CASE 2. $j < i$

In this case, we have $\frac{\partial J_j^{\infty}}{\partial a_{\lambda n}} = 0$ for $\lambda = n - d' + i, \ldots, n - 1$. To see that this is true, note that as in Case 1, we have $\widehat{v}_{\lambda}(P_{\infty,j}) \frac{\partial J_j^{\infty}}{\partial a_{\lambda n}} = \frac{\partial J_j^{\infty}}{\partial a_{\lambda\lambda}}$. If $\lambda > n - d' + j$, we have $\frac{\partial J_j^{\infty}}{\partial a_{\lambda\lambda}} = 0$ by Proposition 5.19. From Corollary 5.14, since $\widehat{v}_{\lambda}(P_{\infty,j}) \neq 0$ on the open dense set of L satisfying $\sum_{\mu=1}^{d'-1} \frac{\partial I_{n_1'\mu,-k_1'\mu}}{\partial a_{n-d'+j,\lambda}}(L)(\sigma_j')^{d'-1-\mu} \neq 0$, $\lambda = n - d' + j$, $\ldots$, n, we infer that $\frac{\partial J_j^{\infty}}{\partial a_{\lambda n}} = 0$ for $\lambda > n - d' + j$ on this open dense set and so it holds in general by continuity. Since $n-d'+j+1 \leq n-d'+i$, our assertion follows. Consequently, it follows from (5.139) that $\{\log \widehat{v}_{n-d'+i}(P_{\infty,i}), J_j^{\infty}\}(L) = 0$.

CASE 3. $i < j \leq d' - 1$

By continuity, it suffices to compute the Poisson bracket on the open dense set of L satisfying the additional condition that $\widehat{v}'(P_{\infty,j})$ exists. For this purpose, consider the sum $\sum_{\lambda=n-d'+i}^{n-1} \widehat{v}_{\lambda}(P_{\infty,i})\Delta_{\lambda n}(P)$. For P on the curve in a neighborhood of $P_{\infty,j}$

(since $i < j$, we have $P_{\infty,i} \neq P_{\infty,j}$), we introduce a parameter t as before. Thus $P = P(t) = (z(t), w(t))$, where

$$z(t) = c'_j t^{-n'_1} + \cdots, \quad w(t) = t^{-(n'_1+k'_1)}, \quad \text{and } (c'_j)^{n'_1+k'_1} = \frac{1}{\sigma'_j(L)}. \tag{5.145}$$

By using the asymptotics in Proposition 5.13 (c), we have

$$\begin{aligned} &\sum_{\lambda=n-d'+i}^{n-1} \widehat{v}_\lambda(P_{\infty,i})\Delta_{\lambda n}(P(t)) \\ =&(\sigma'_j)^{-(d'-1)} \left(\sum_{\lambda=n-d'+i}^{n-1} \widehat{v}_\lambda(P_{\infty,i}) \frac{\partial J_j^\infty}{\partial a_{\lambda n}} \right) t^{-(n'_1+k'_1)(n-1)} + O(t^{-(n'_1+k'_1)(n-1)+1}) \\ &\text{as } t \to 0 \end{aligned} \tag{5.146}$$

from which we obtain the representation

$$\sum_{\lambda=n-d'+i}^{n-1} \widehat{v}_\lambda(P_{\infty,i}) \frac{\partial J_j^\infty}{\partial a_{\lambda n}} = \lim_{t\to 0} \frac{(\sigma'_j)^{d'-1} \sum_{\lambda=n-d'+i}^{n-1} \widehat{v}_\lambda(P_{\infty,i})\Delta_{\lambda n}(P(t))}{t^{-(n'_1+k'_1)(n-1)}}. \tag{5.147}$$

Our next task is to evaluate the limit on the right hand side of the above expression. Since $\widehat{v}_\lambda(P_{\infty,i}) = 0$ for $\lambda = 1, \ldots, n - d' + i - 1$, we have

$$\begin{aligned} \lim_{t\to 0} \frac{\sum_{\lambda=n-d'+i}^{n-1} \widehat{v}_\lambda(P_{\infty,i})\Delta_{\lambda n}(P(t))}{t^{-(n'_1+k'_1)(n-1)}} &= \lim_{t\to 0} \frac{\sum_{\lambda=1}^{n} \widehat{v}_\lambda(P_{0,i})\Delta_{\lambda n}(P(t)) - \Delta_{nn}(P(t))}{t^{-(n'_1+k'_1)(n-1)}} \\ &= \lim_{t\to 0} \frac{\left(\sum_{\lambda=1}^{n} \widehat{v}_\lambda(P_{\infty,i})\widehat{v}'_\lambda(P(t))\right) \Delta_{1n}(P(t))}{t^{-(n'_1+k'_1)(n-1)}}, \end{aligned} \tag{5.148}$$

where we have used the fact that $\Delta_{nn}(P(t)) = O(t^{-(n'_1+k'_1)(n-1)+1})$ as $t \to 0$. (This is a consequence of Proposition 5.16.) By Proposition 5.13 (d), we have

$$\lim_{t\to 0} \frac{\Delta_{1n}(P(t))}{t^{-(n'_1+k'_1)(n-1)-(n'_1-1)}} = \sum_{\substack{(r,s) \\ k'_1 r + n'_1 s = 1}} (c'_j)^{r+1-s} \frac{\partial I_{r,s}}{\partial a_{1n}}. \tag{5.149}$$

Consequently, and in view of the factorization

$$\begin{aligned} &\frac{\left(\sum_{\lambda=1}^{n} \widehat{v}_\lambda(P_{\infty,i})\widehat{v}'_\lambda(P(t))\right) \Delta_{1n}(P(t))}{t^{-(n'_1+k'_1)(n-1)}} \\ =&\frac{\sum_{\lambda=1}^{n} \widetilde{v}_\lambda(P_{\infty,i})\widehat{v}'_\lambda(P(t))}{t^{n'_1-1}} \, \frac{\Delta_{1n}(P(t))}{t^{-(n'_1+k'_1)(n-1)-(n'_1-1)}}, \end{aligned} \tag{5.150}$$

it follows that in order to compute the right hand side of (5.147), it remains to compute

$$\lim_{t\to 0} \frac{\sum_{\lambda=1}^{n} \widehat{v}_\lambda(P_{\infty,i})\widehat{v}'_\lambda(P(t))}{t^{n'_1-1}}. \tag{5.151}$$

Note that as a consequence of biorthogonality, we have $\sum_{\lambda=1}^{n} \widehat{v}_\lambda(P_{\infty,i})\widehat{v}'_\lambda(P_{\infty,j}) = 0$. Therefore the limit in (5.151) is that of an indeterminate form. We will evaluate this limit by computing

$$\lim_{t\to 0} \frac{\sum_{\lambda=1}^{n} \widehat{v}_\lambda(P_{\infty,i}) \frac{d}{dt}\widehat{v}'_\lambda(P(t))}{(n'_1 - 1)t^{n'_1-2}} \tag{5.152}$$

and appeal to L'Hôpital's rule. To this end, we will need to use the formula

$$(L(z(t))^{n_1'}z(t)^{-k_1'})^T\widehat{v}'(P(t)) = w(t)^{n_1'}z(t)^{-(n_1'+k_1')}\widehat{v}'(P(t)). \tag{5.153}$$

We begin by differentiating both sides of (5.153) with respect to t, divide the result by $t^{n_1'-2}$, and pair it with $\widehat{v}(P_{\infty,i})$, this results in

$$\begin{aligned}&\left(\widehat{v}(P_{\infty,i}), \frac{\frac{d}{dt}(L(z(t))^{n_1'}z(t)^{-k_1'})^T}{t^{n_1'-2}}\widehat{v}'(P(t)\right)\\ &\quad+\left(L(z(t))^{n_1'}z(t)^{-k_1'}\widehat{v}(P_{\infty,i}), \frac{\frac{d}{dt}\widehat{v}'(P(t))}{t^{n_1'-2}}\right)\\ &=\frac{d}{dt}a(t)\left(\widehat{v}(P_{\infty,i}), \frac{\widehat{v}'(P(t))}{t^{n_1'-2}}\right)+a(t)\left(\widehat{v}(P_{\infty,i}), \frac{\frac{d}{dt}\widehat{v}'(P(t))}{t^{n_1'-2}}\right),\end{aligned} \tag{5.154}$$

where $a(t) = w(t)^{n_1'}z(t)^{-(n_1'+k_1')}$. We will analyze the limiting behaviour of each of the terms in the above expression as $t \to 0$. First of all, from the divisor $(\widehat{v}_\lambda)$ (see (5.50)), we have

$$\widehat{v}_\lambda(P(t)) = O(t^{n_1'-1}) \text{ as } t \to 0, \quad n - d' + 1 \leq \lambda \leq n - 1 \tag{5.155}$$

from which we deduce that

$$\lim_{t\to 0}\left(\widehat{v}(P_{\infty,i}), \frac{\frac{d}{dt}\widehat{v}'(P(t))}{t^{n_1'-2}}\right) \text{ exists.} \tag{5.156}$$

As $a(t) = \sigma_j'(L) + O(t)$ as $t \to 0$, it also follows from (5.155) that

$$\lim_{t\to 0}\frac{d}{dt}a(t)\left(\widehat{v}(P_{\infty,i}), \frac{\widehat{v}'(P(t))}{t^{n_1'-2}}\right) = 0. \tag{5.157}$$

On the other hand, we obtain from Proposition 2.3 and Corollary 2.4 that

$$L(z(t))^{n_1'}z(t)^{-k_1'} = \lim_{\tau\to 0} L(z(\tau))^{n_1'}z(\tau)^{-k_1'} + O(t^{n_1'}), \quad t \to 0. \tag{5.158}$$

Hence

$$\lim_{t\to 0}\left(\widehat{v}(P_{\infty,i}), \frac{\frac{d}{dt}(L(z(t))^{n_1'}z(t)^{-k_1'})^T}{t^{n_1'-2}}\widehat{v}'(P(t))\right) = 0. \tag{5.159}$$

Since

$$\lim_{t\to 0}\left(L(z(t))^{n_1'}z(t)^{-k_1'}\widehat{v}(P_{\infty,i}), \frac{\frac{d}{dt}\widehat{v}'(P(t))}{t^{n_1'-2}}\right) = \sigma_i'(L)\lim_{t\to 0}\left(\widehat{v}(P_{\infty,i}), \frac{\frac{d}{dt}\widehat{v}'(P(t))}{t^{n_1'-2}}\right), \tag{5.160}$$

upon taking the limit as $t \to 0$ on both sides of (5.154), it follows on using our analysis above that

$$\sigma_i'(L)\lim_{t\to 0}\left(\widehat{v}(P_{\infty,i}), \frac{\frac{d}{dt}\widehat{v}'(P(t))}{t^{n_1'-2}}\right) = \sigma_j'(L)\lim_{t\to 0}\left(\widehat{v}(P_{\infty,i}), \frac{\frac{d}{dt}\widehat{v}'(P(t))}{t^{n_1'-2}}\right) \tag{5.161}$$

and hence (as $\sigma_i'(L) \neq \sigma_j'(L)$)

$$\lim_{t\to 0}\left(\widehat{v}(P_{\infty,i}), \frac{\frac{d}{dt}\widehat{v}'(P(t))}{t^{n_1'-2}}\right) = 0. \tag{5.162}$$

Accordingly,

$$\sum_{\lambda=n-d'+i}^{n-1} \widehat{v}_\lambda(P_{\infty,i})\frac{\partial J_j^\infty}{\partial a_{\lambda n}} = 0 \tag{5.163}$$

and so $\{\log \widehat{v}_i(P_{\infty,i}), J_j^\infty\}(L) = 0$. □

By using Theorem 5.12, Theorem 5.21, and Theorem 4.3, we are now ready to prove the next result.

THEOREM 5.22. *The periodic Toda flow with Hamiltonian $H(L) = \frac{1}{2}(L, L)_{L\mathfrak{g}}$ is Liouville integrable on the generic coadjoint orbits in $\mathcal{T}^{(k,k')}$ for all (k, k') satisfying $1 \leq k, k' \leq n-1$, $k + k' < n$.*

PROOF. We have to establish the functional independence of the nontrivial integrals on an open, dense subset of each generic coadjoint orbit. So let $L \in \mathcal{G}$ satisfy the additional genericity assumptions and suppose

$$\begin{aligned} 0 = &\sum_{r=1}^{n-1} \sum_{j=-\lfloor \frac{k'd_r}{h} \rfloor+1}^{\lfloor \frac{kd_r}{h} \rfloor -1} c_{rj} dI_{r,j}(L) + \sum_{r \notin E} c_{r,\lfloor \frac{kd_r}{h} \rfloor} dI_{r,\lfloor \frac{kd_r}{h} \rfloor}(L) \\ &+ \sum_{\substack{r \notin E' \\ n_1' \nmid r}} c_{r,-\lfloor \frac{k'd_r}{h} \rfloor} dI_{r,-\lfloor \frac{k'd_r}{h} \rfloor}(L) + \sum_{j=1}^{d'-1} c_j dJ_j^\infty(L). \end{aligned} \tag{5.164}$$

If $X_{\phi_{ab}}$ denote the Hamiltonian vector field generated by ϕ_{ab}, then by applying both sides of (5.164) to $-X_{\phi_{ab}}(L)$, it follows from Theorem 5.12 that

$$\begin{aligned} 0 = &\sum_{r=1}^{n-1} \sum_{j=-\lfloor \frac{k'd_r}{h} \rfloor+1}^{\lfloor \frac{kd_r}{h} \rfloor -1} c_{rj} \{\phi_{ab}, I_{r,j}\}(L) + \sum_{r \notin E} c_{r,\lfloor \frac{kd_r}{h} \rfloor} \left\{\phi_{ab}, I_{r,\lfloor \frac{kd_r}{h} \rfloor}\right\}(L) \\ &+ \sum_{\substack{r \notin E' \\ n_1' \nmid r}} c_{r,-\lfloor \frac{k'd_r}{h} \rfloor} \left\{\phi_{ab}, I_{r,-\lfloor \frac{k'd_r}{h} \rfloor}\right\}(L) \\ = &\sum_{r=1}^{n-1} \sum_{j=-\lfloor \frac{k'd_r}{h} \rfloor+1}^{\lfloor \frac{kd_r}{h} \rfloor -1} c_{rj} \delta_{r,n-b}\delta_{j,n-a-b} + \sum_{r \notin E} c_{r,\lfloor \frac{kd_r}{h} \rfloor} \delta_{r,n-b}\delta_{\lfloor \frac{kd_r}{h} \rfloor,n-a-b} \\ &+ \sum_{\substack{r \notin E' \\ n_1' \nmid r}} c_{r,-\lfloor \frac{k'd_r}{h} \rfloor} \delta_{r,n-b}\delta_{-\lfloor \frac{k'd_r}{h} \rfloor,n-a-b} \\ = &\, c_{n-b,n-a-b}. \end{aligned} \tag{5.165}$$

Since the map $(r, j) \mapsto (a, b) = (r - j, n - r)$ is a bijection between the union of the following four sets

(5.166)

$$\left\{(r,j) \mid -\lfloor \frac{k'd_r}{h} \rfloor + 1 \leq j \leq \lfloor \frac{kd_r}{h} \rfloor - 1, 1 \leq r \leq n-1\right\}, \left\{\left(r, \lfloor \frac{kd_r}{h} \rfloor\right) \mid r \notin E\right\},$$
$$\left\{\left(r, -\lfloor \frac{k'd_r}{h} \rfloor\right) \mid r \notin E', n_1' \nmid r\right\}, \{(n_1'j, -k_1'j) \mid j = 1, \ldots, d'-1\}$$

and $\Delta(I)^o \cap \mathbb{N}$, it follows that (5.164) reduces to

$$0 = \sum_{j=1}^{d'-1} c_j dJ_j^{\infty}(L). \tag{5.167}$$

Next, let $X_{\log \widehat{v}_{n-d'+i}(P_{\infty,i})}$ be the Hamiltonian vector field generated by

$$\log \widehat{v}_{n-d'+i}(P_{\infty,i}).$$

If we apply both sides of (5.167) to $-X_{\log \widehat{v}_{n-d'+i}(P_{\infty,i})}$, it follows from Theorem 5.21 that

$$\begin{aligned} 0 &= \sum_{j=1}^{d'-1} c_j \{\log \widehat{v}_{n-d'+i}(P_{\infty,i}), J_j^{\infty}\}(L) \\ &= \frac{\partial J_i^{\infty}}{\partial a_{n-d'+i,n-d'+i}} c_i \end{aligned} \tag{5.168}$$

and hence $c_i = 0$ as well. □

CHAPTER 6

Correspondence between $\mathcal{T}_*^{(k,k')}$ and Extended Algebro-Geometric Data

We begin by defining the subset $\mathcal{T}_*^{(k,k')} \subset \mathcal{T}^{(k,k')}$ which is the domain of the map into the collection of extended algebro-geometric data. This consists essentially of putting together all the assumptions we have made in the various steps of our analysis in the previous chapters.

DEFINITION 6.1. The open subset $\mathcal{T}_*^{(k,k')} \subset \mathcal{T}^{(k,k')}$ consists of those $L \in \mathcal{T}^{(k,k')}$ satisfying the following conditions:

(a) $L \in \mathcal{G}$,
(b) the polynomial $I(L,z,w)$ is irreducible,
(c) if $\bar{I}_1(L,z)$, $\bar{I}_2(L,z)$ and $\bar{I}_3(L,z)$ are defined by $I(L,z,0) = z^{n-k}\bar{I}_1(L,z)$, $\frac{\partial I}{\partial z}(L,z,0) = z^{n-k-1}\bar{I}_2(L,z)$, $\frac{\partial I}{\partial w}(L,z,0) = z^{n-k}\bar{I}_3(L,z)$, and $R_{\alpha,\beta}(L)$ are defined by (5.26), then $R_{\alpha,\beta}(L) \neq 0$ for all α, β,
(d) the resultant $\operatorname{Res}(\bar{I}_1(L,\cdot),\bar{I}_2(L,\cdot)) \neq 0$, or equivalently, the map

$$\pi : C(L) \longrightarrow \mathbb{CP}^1, [\xi : z : w] \mapsto [\xi : w] \tag{6.1}$$

has no ramification points of the form $[1 : z : 0]$,
(e) $\Delta_A(I(L,\cdot,\cdot)) \neq 0$, where Δ_A is the A-resultant for A given by the set of lattice points in the triangular region with vertices at $(0,n)$, $(n-k,0)$ and $(n+k',0)$.
(f) $\widehat{v}_{n-d'+i}(P_{\infty,i})$ exists and is nonzero, $i = 1, \ldots, d'$. (See Corollary 5.14.)

DEFINITION 6.2. The *extended algebro-geometric data* associated to $L \in \mathcal{T}_*^{(k,k')}$ consists of:

(a) the spectral curve $C(L)$ with normalization $\phi : \widetilde{C}(L) \longrightarrow C(L)$,
(b) the divisor $D(L)$,
(c) the component $\widehat{v}_{n-d'+i}(P_{\infty,i})$ of the eigenvector $\widehat{v}(P_{\infty,i})$, $i = 1, \ldots, d'$,
(d) the coadjoint orbit invariants given by the entries $a_{1,k'+1}, \ldots, a_{n-k',n}$ on the k'-th superdiagonal of L_0, and the entries $a_{n-k'+1,1}, \ldots, a_{n-d',k'-d'}$ on the $(n-k')$-th subdiagonal of L_1.

In this chapter, our goal is to study the correspondence

$$\mathcal{T}_*^{(k,k')} \ni L \mapsto \text{the extended algebro-geometric data associated to } L. \tag{6.2}$$

This will consist of two parts, the forward problem and the inverse problem.

6.1. The forward problem

THEOREM 6.3. *The loop $L \in \mathcal{T}_*^{(k,k')}$ is uniquely determined by its associated extended algebro-geometric data. That is, the map*

(6.3)
$$\begin{aligned}\mathcal{T}_*^{(k,k')} \ni L \mapsto & \text{the spectral curve } C(L) \text{ with normalization } \phi : \widetilde{C}(L) \longrightarrow C(L),\\ & \text{the divisor } D(L), (\widehat{v}_{n-d'+1}(P_{\infty,1}), \dots, \widehat{v}_n(P_{\infty,d'})),\\ & (a_{1,k'+1}, \dots, a_{n-k',n}), (a_{n-k'+1,1}, \dots, a_{n-d',k'-d'})\end{aligned}$$

is one-to-one.

PROOF. We will show, successively, that $\sigma_i'(L)$, $1 \le i \le d'$, the boundary elements $a_{n-k'+i,i}$, $k'-d'+1 \le i \le n$, $\widehat{v}(P)$, and $L(z)$ are uniquely determined by the associated extended algebro-geometric data. The way we accomplish this is by deriving formulas for these quantities in terms of the extended algebro-geometric data.

STEP 1. From the curve $C(L)$, it follows that

$$\begin{aligned} I_{T3}(z,w) :=& (-1)^n w^n + \sum_{j=1}^{d'} I_{n_1'j-1,-k_1'j}(L) z^{(n_1'+k_1')j} w^{n_1'(d'-j)} \\ =& (-1)^n \prod_{i=1}^{d'} (w^{n_1'} - \sigma_i'(L) z^{n_1'+k_1'}). \end{aligned} \tag{6.4}$$

Hence $\sigma_1'(L), \dots, \sigma_{d'}'(L)$ are determined by $I(L,z,w)$ only up to some order. From this, we see that the points $P_{\infty,1}, \dots, P_{\infty,d'}$ are really necessary to fix the $\sigma_i'(L)$'s. This observation will be important in the next section when we formulate the notion of an admissible set of data.

Now by counting the number of variables from L, we see that the dimension of the domain of the map is equal to $n(k+k')+n$. On the other hand, since $\widehat{v}(P_{\infty,d'}) = 1$, it follows by counting the number of $I_{r,j}$ and the genus formula that the dimension of the codomain is also given by the same number.

STEP 2. Now, from $a_{i,k'+i}$, $1 \le i \le n-k'$, $a_{n-k'+i,i}$, $1 \le i \le k'-d'$, we can construct

(6.5)
$$\tau_i'(L) = \prod_{j=1}^{n_1'-k_1'} a_{i+(j-1)d',k'+i+(j-1)d'} \prod_{j=1}^{k_1'-1} a_{n-k'+i+(j-1)d',i+(j-1)d'}, 1 \le i \le d'.$$

Hence it follows that

$$a_{n-k'+i+(k_1'-1)d',i+(k_1'-1)d'} = \frac{\sigma_i'(L)}{\tau_i'(L)}, \quad 1 \le i \le d'. \tag{6.6}$$

STEP 3. Fix a canonical homology basis $\{a_i, b_j\}_{1 \le i,j \le g}$ of the normalization $\widetilde{C}(L)$ of the spectral curve $C(L)$ and let $\{\omega_i\}_{1 \le i \le g}$ be a cohomology basis dual to the above homology basis, i.e.,

$$\int_{a_j} \omega_i = \delta_{ij}, \quad \int_{b_j} \omega_i = \Omega_{ij}. \tag{6.7}$$

With respect to the Riemann matrix $\Omega = (\Omega_{ij})$, construct the theta function

$$\theta(z) = \theta(z,\Omega) = \sum_{n\in\mathbb{Z}^g} \exp\{2\pi i(n,z) + \pi i(n,\Omega n)\}. \tag{6.8}$$

Let $\omega = (\omega_1,\dots,\omega_g)$ and choose a non-singular $e \in \mathbb{C}^g$ in the theta divisor, i.e. $\theta(e) = 0$ and such that the prime form $E_e(x,y) \equiv \theta\left(e + \int_x^y \omega\right) \not\equiv 0$, with the additional property that $E_e(P_{\infty,i},P) \not\equiv 0$ for each $1 \le i \le d'$. The existence of an e is guaranteed by Lemma 3.3 of [**Mum**]. Hence by Corollary 3.6 of [**Mum**], there exists $D_{g-1} \in \mathrm{Div}^{g-1}(\widetilde{C}(L))$ such that

$$e = \Delta - \int_{(g-1)P_{0,1}}^{D_{g-1}} \omega \tag{6.9}$$

where Δ is the vector of Riemann constants. Note that we have
(6.10)

$$\theta\left(e + \int_{D(L)}^{D_{g-1}+P} \omega\right) = \theta\left(\Delta + \int_{D(L)}^{(g-1)P_{0,1}+P} \omega\right) = 0 \iff P \in \operatorname{supp} D(L).$$

Therefore, according to (5.48), we have

$$f_j = \widetilde{c}_j \frac{\prod_{b=0}^{j-1} E_e(P_{0,b},P)}{\prod_{a=1}^{j} E_e(P_{\infty,a},P)} \frac{\theta\left(e + \int_{D(L)+\sum_{a=1}^{j} P_{\infty,a}}^{D_{g-1}+\sum_{b=0}^{j-1} P_{0,b}+P} \omega\right)}{\theta\left(e + \int_{D(L)}^{D_{g-1}+P} \omega\right)}. \tag{6.11}$$

Since

$$z = \text{constant}\, \frac{\prod_{b=1}^{d} E_e(P_{0,b},P)^{n_1}}{\prod_{a=1}^{d'} E_e(P_{\infty,a},P)^{n'_1}}, \tag{6.12}$$

it follows that

$$\widehat{v}_j(P) = c_j \frac{\prod_{a=j+1}^{n} E_e(P_{\infty,a},P)}{\prod_{b=j}^{n-1} E_e(P_{0,b},P)} \frac{\theta\left(e + \int_{D(L)+\sum_{a=1}^{j} P_{\infty,a}}^{D_{g-1}+\sum_{b=0}^{j-1} P_{0,b}+P} \omega\right)}{\theta\left(e + \int_{D(L)}^{D_{g-1}+P} \omega\right)}. \tag{6.13}$$

Hence we can write

$$\widehat{v}(P) = c\widehat{v}^{\theta}(P) \tag{6.14}$$

where $\widehat{v}^{\theta}(P)$ is known and $c = \mathrm{diag}(c_1,\dots,c_n)$ is to be determined. Note that the above gives, in particular, that
(6.15)

$$\widehat{v}_{n-d'+i}(P) = c_{n-d'+i} \frac{\prod_{a=i+1}^{d'} E_e(P_{\infty,a},P)}{\prod_{b=n-d'+i}^{n-1} E_e(P_{0,b},P)} \frac{\theta\left(e + \int_{D(L)+\sum_{a=1}^{n-d'+i} P_{\infty,a}}^{D_{g-1}+\sum_{b=0}^{n-d'+i-1} P_{0,b}+P} \omega\right)}{\theta\left(e + \int_{D(L)}^{D_{g-1}+P} \omega\right)}.$$

Consequently, for $1 \le i \le d'$, the constants $c_{n-d'+i}$ are determined from the equations

$$\widehat{v}_{n-d'+i}(P_{\infty,i}) = c_{n-d'+i}\frac{\prod_{a=i+1}^{d'} E_e(P_{\infty,a},P_{\infty,i})}{\prod_{b=n-d'+i}^{n-1} E_e(P_{0,b},P_{\infty,i})}\frac{\theta\left(e+\int_{D(L)+\sum_{a=1}^{n-d'+i}P_{\infty,a}}^{D_{g-1}+\sum_{b=0}^{n-d'+i-1}P_{0,b}+P_{\infty,i}}\omega\right)}{\theta\left(e+\int_{D(L)}^{D_{g-1}+P_{\infty,i}}\omega\right)}. \tag{6.16}$$

Incidentally, note that it follows from the above expression in (6.16) and the fact in (6.10) that $P_{\infty,i}\notin \operatorname{supp} D(L)$, $i=1,\ldots,d'$. This fact (which we obtained already in Corollary 5.14), will play a role when we formulate the notion of an admissible set of data in the next subsection.

STEP 4. We next show how to determine $c_1,\ldots,c_{n-d'}$. For given $z\in\mathbb{CP}^1$ which is not a branch point of $C(L)$, let $P_1(z)=[1:z:w_1(z)],\ldots,P_n(z)=[1:z:w_n(z)]$ be the n points on $C(L)$ lying over z. Let

$$\begin{aligned}\widehat{V}(z)&=(\widehat{v}(P_1(z)),\ldots,\widehat{v}(P_n(z))),\\ \widehat{V}^\theta(z)&=(\widehat{v}^\theta(P_1(z)),\ldots,\widehat{v}^\theta(P_n(z))).\end{aligned}\tag{6.17}$$

With these definitions, it follows from (6.14) that

$$\widehat{V}(z)=c\,\widehat{V}^\theta(z) \tag{6.18}$$

and so

$$\begin{aligned} zL(z) =&\widehat{V}(z)\Lambda(z)\widehat{V}(z)^{-1},\quad \Lambda(z)=\operatorname{diag}(w_1(z),\ldots,w_n(z)),\\ =&c\,\widehat{V}^\theta(z)\Lambda(z)\widehat{V}^\theta(z)^{-1}c^{-1}\end{aligned}\tag{6.19}$$

where $c_1,\ldots,c_{n-d'}$ are as yet to be determined. To simplify notation, let $z\widetilde{B}(z)=B_1z^2+B_0z+B_{-1}=\widehat{V}^\theta(z)\Lambda(z)\widehat{V}^\theta(z)^{-1}$. We want to compare the k'-th superdiagonal and the $(n-k')$-th subdiagonal of the matrices on both sides of the equation $L(z)\,c=c\,\widetilde{B}(z)$. In what follows, it is convenient to partition our matrices into $d'\times d'$ blocks. For this purpose, let

$$\gamma_1=\operatorname{diag}(c_1,\ldots,c_{d'}),\ldots,\gamma_{n_1'}=\operatorname{diag}(c_{n-d'+1},\ldots,c_n). \tag{6.20}$$

Taking the diagonal of the $d'\times d'$ blocks on the k_1'-th block superdiagonal of L_0 and B_0, we obtain

$$\begin{pmatrix} D_1 & & \\ & \ddots & \\ & & D_{n_1'-k_1'}\end{pmatrix},\begin{pmatrix} E_1 & & \\ & \ddots & \\ & & E_{n_1'-k_1'}\end{pmatrix}, \tag{6.21}$$

where the diagonal matrices are given by

$$
\begin{aligned}
&D_1 = \operatorname{diag}(a_{1,k'+1}, \dots, a_{d',k'+d'}),\\
&\vdots\\
&D_{n_1'-k_1'} = \operatorname{diag}(a_{n-k'-d'+1,n-d'+1}, \dots, a_{n-k',n}),\\
&E_1 = \operatorname{diag}(b_{1,k'+1}, \dots, b_{d',k'+d'}),\\
&\vdots\\
&E_{n_1'-k_1'} = \operatorname{diag}(b_{n-k'-d'+1,n-d'+1}, \dots, b_{n-k',n}).
\end{aligned}
\tag{6.22}
$$

Similarly, by taking the diagonal of the $d' \times d'$ blocks on the $(n_1' - k_1')$-th block subdiagonal of L_1 and B_1, we find

$$
\begin{pmatrix} D_{n_1'-k_1'+1} & & \\ & \ddots & \\ & & D_{n_1'} \end{pmatrix}, \begin{pmatrix} E_{n_1'-k_1'+1} & & \\ & \ddots & \\ & & E_{n_1'} \end{pmatrix},
\tag{6.23}
$$

where the diagonal matrices are

$$
\begin{aligned}
&D_{n_1'-k_1'+1} = \operatorname{diag}(a_{n-k'+1,1}, \dots, a_{n-k'+d',d'}),\\
&\vdots\\
&D_{n_1'} = \operatorname{diag}(a_{n-d'+1,k'-d'+1}, \dots, a_{nk'}),\\
&E_{n_1'-k_1'+1} = \operatorname{diag}(b_{n-k'+1,1}, \dots, b_{n-k'+d',d'}),\\
&\vdots\\
&E_{n_1'} = \operatorname{diag}(b_{n-d'+1,k'-d'+1}, \dots, b_{nk'}).
\end{aligned}
\tag{6.24}
$$

Therefore, when we equate the diagonal of the $d' \times d'$ blocks on the k_1'-th block superdiagonal (resp. $(n_1' - k_1')$-th block subdiagonal) of the matrices on both sides of the equation $zL(z)c = c\widetilde{B}(z)$, we obtain the equations

$$
\begin{aligned}
&\gamma_i E_i = D_i \gamma_{k_1'+i}, \quad i = 1, \dots, n_1' - k_1',\\
&\gamma_{n_1'-k_1'+i} E_{n_1'-k_1'+i} = D_{n_1'-k_1'+i}\, \gamma_i, \quad i = 1, \dots, k_1'
\end{aligned}
\tag{6.25}
$$

which can be rewritten as

$$
\gamma_i E_i = D_i\, \gamma_{\overline{\sigma}(i)}, \quad i = 1, \dots, n_1',
\tag{6.26}
$$

where $\overline{\sigma}$ is the permutation of $\{1, \dots, n_1'\}$ defined by

$$
\overline{\sigma} = \begin{pmatrix} 1 & \cdots & n_1' - k_1' & n_1' - k_1' + 1 & \cdots & n_1' \\ k_1' + 1 & \cdots & n_1' & 1 & \cdots & k_1' \end{pmatrix},
\tag{6.27}
$$

and where $\gamma_{n_1'}$ has been determined above in (6.16).

Next, we make the observation that if we define s such that

$$
s(i) = \overline{\sigma}^i(n_1'), \quad i = 1, \dots, n_1',
\tag{6.28}
$$

then s is a cyclic permutation of $\{1, \dots, n_1'\}$. To see that this is true, note that $\overline{\sigma}(j) \equiv j + k_1' \pmod{n_1'}$ from which we find $s(i) \equiv ik_1' \pmod{n_1'}$. As $\gcd(n_1', k_1') = 1$, it follows that the numbers $ik_1' \pmod{n_1'}$, $i = 1, \dots, n_1'$ are all distinct and

$s(n_1') = n_1'$. Consequently, starting from $\gamma_{n_1'}$, we can solve for $\gamma_{s(1)}, \gamma_{s(2)}, \dots, \gamma_{s(n_1'-1)}$ successively from (6.26). Since c is now completely determined, we can insert it back in (6.14) and (6.19). This demonstrates that $\widehat{v}(P)$ and $L(z)$ are uniquely determined by the associated extended algebro-geometric data. □

6.2. The inverse problem

In this section, our goal is to solve the inverse problem, which is the problem of characterizing the collection of extended algebro-geometric data.

DEFINITION 6.4. An admissible set of data consists of

(a) a curve C in $\mathbb{P}^2(1,1,2)$ with two singular points $P_0 = [1:0:0]$, $P_\infty = [0:1:0]$ whose affine part is defined by the irreducible polynomial

$$I(z,w) = (-1)^n w^n + \sum_{r=0}^{n-1} \left(\sum_{j=-\lfloor \frac{k'd_r}{h} \rfloor}^{\lfloor \frac{kd_r}{h} \rfloor} I_{r,j} z^{r+1-j} \right) w^{n-1-r} \tag{6.29}$$

and whose normalization $\phi : \widetilde{C} \longrightarrow C$ is such that

$$\begin{aligned} \phi^{-1}(P_0) &= \{P_{0,1}, \dots, P_{0,d}\}, \\ \phi^{-1}(P_\infty) &= \{P_{\infty,1}, \dots, P_{\infty,d'}\}, \end{aligned} \tag{6.30}$$

where

$$\begin{aligned} P_{0,i} &= \langle c_i t^{n_1} + \cdots, t^{n_1-k_1} \rangle, \quad c_i^{n_1-k_1} = \frac{1}{\sigma_i}, 1 \le i \le d, \\ P_{\infty,j} &= \langle c_j' t^{-n_1'} + \cdots, t^{-(n_1'+k_1')} \rangle, \quad (c_j')^{(n_1'+k_1')} = \frac{1}{\sigma_j'}, 1 \le j \le d'. \end{aligned} \tag{6.31}$$

Here $\langle f(t), g(t) \rangle$ for a pair $(f(t), g(t))$ of meromorphic functions on an open neighborhood of $0 \in \mathbb{C}$ is the equivalence class containing (f,g) where the equivalence relation is defined by (see [**Kir**] for the approach on normalization that we are using here)

$$\begin{aligned} (f,g) \sim (\widetilde{f}, \widetilde{g}) \iff & \text{there is a holomorphic parameter change } \rho \text{ satisfying} \\ & \rho(0) = 0, \rho'(0) \neq 0 \text{ such that } \widetilde{f} = f \circ \rho, \widetilde{g} = g \circ \rho \\ & \text{in some open neighborhood of } 0, \end{aligned} \tag{6.32}$$

(b) a divisor D of degree equal to the genus of $\widetilde{C}$ which is *general*, and is *regular* with respect to $\{P_{0,i}\}_{i\in\mathbb{Z}}$ and $\{P_{\infty,j}\}_{j\in\mathbb{Z}}$ ($P_{0,i+d} = P_{0,i}$ and $P_{\infty,i+d'} = P_{\infty,i}$ for all i) in the sense that for all j, we have

$$\dim \mathscr{L}\left(D + \sum_{i=1}^{j} P_{\infty,j} - \sum_{i=0}^{j} P_{0,i} \right) = 0, \tag{6.33}$$

and moreover, $\operatorname{supp} D \cap \{P_{\infty,1}, \dots, P_{\infty,d'}\} = \emptyset$,

(c) $(v_{n-d'+1}, \dots, v_{n-1}, v_n = 1) \in (\mathbb{C}^*)^{d'}$,

(d) $(a_{1,k'+1}, \dots, a_{n-k',n}) \in (\mathbb{C}^*)^{n-k'}$, $(a_{n-k'+1,1}, \dots, a_{n-d',k'-d'}) \in (\mathbb{C}^*)^{k'-d'}$.

We will denote the collection of all admissible sets of data by $\mathcal{D}$.

REMARK 6.5.
(a) Note that in (6.33), the sum $\sum_{i=1}^{j} P_{\infty,i}$ is equal to 0 for $j = 0$, and is given by $-P_{\infty,0} - P_{\infty,-1} - \cdots - P_{\infty,j+1}$ for $j \le -1$ [**vMM**]. Similarly, the sum $\sum_{i=0}^{j} P_{0,i}$ is equal to 0 for $j = -1$ and is given by $-P_{0,-1} - P_{\infty,-2} - \cdots - P_{\infty,j+1}$ for $j \le -2$.
(b) The notion of regularity of a divisor in Definition 6.4 (b) above is due to van Moerbeke and Mumford in [**vMM**].
(c) Note that the σ_i are nonzero and are distinct and the same holds for the σ'_j.

Our theorem is the following.

THEOREM 6.6. *The map*

(6.34)

$$\begin{aligned}\mathcal{T}_*^{(k,k')} \ni L \mapsto &\text{ the spectral curve } C(L) \text{ with normalization } \phi : \widetilde{C}(L) \longrightarrow C(L),\\ &\text{the divisor } D(L), (\widehat{v}_{n-d'+1}(P_{\infty,1}), \ldots, \widehat{v}_n(P_{\infty,d'})),\\ &(a_{1,k'+1}, \ldots, a_{n-k',n}), (a_{n-k'+1,1}, \ldots, a_{n-d',k'-d'})\end{aligned}$$

is a bijection onto $\mathcal{D}$.

PROOF. Our discussion in Section 5 shows that for $L \in \mathcal{T}_*^{(k,k')}$, the curve $C(L)$ has the properties in Definition 6.4 (a). On the other hand, Lemma 4 in [**vMM**] shows that the divisor $D(L)$ is regular with regard to the two sequences $\{P_{0,i}\}_{i\in\mathbb{Z}}$, $\{P_{\infty,i}\}_{i\in\mathbb{Z}}$. By the assumption in Definition 6.1 (f), $\widehat{v}_{n-d'+i}(P_{\infty,i}) \neq 0$, $i = 1, \ldots, d' - 1$. Therefore, from the expression in (6.16) and the fact in (6.10), we see that $\operatorname{supp} D(L) \cap \{P_{\infty,1}, \ldots, P_{\infty,d'}\} = \emptyset$. Moreover, since $L \in \mathcal{G}$, the entries on the k'-th superdiagonal of L_0 and the $(n-k')$-th subdiagonal of L_{-1} are different from 0. Therefore, the image of the map is in $\mathcal{D}$. Furthermore, by Theorem 6.3, the map is one-to-one.

To show that the map is onto, we take an admissible set of data from $\mathcal{D}$ which consist of what we listed in Definition 6.4 above. We will construct $L \in \mathcal{T}_*^{(k,k')}$ such that $I(L, z, w) = I(z, w)$, $D(L) = D$ etc.

STEP 1. Since $P_{\infty,i} = \langle c'_i t^{-n'_1} + \cdots, t^{-(n'_1+k'_1)}\rangle$, when we substitute $z = c'_i t^{-n'_1} + \cdots$, $w = t^{-(n'_1+k'_1)}$ into $I(z, w) = 0$ and equate the lowest order term in t, we find

$$(-1)^n (\sigma'_i)^{d'} + \sum_{j=1}^{d'} I_{n'_1 j-1, -k'_1 j} (\sigma'_i)^{d'-j} = 0. \tag{6.35}$$

Since this is true for all $1 \le i \le d'$, it follows that

$$(-1)^n z^{d'} + \sum_{j=1}^{d'} I_{n'_1 j-1, -k'_1 j} z^{d'-j} = (-1)^n \prod_{i=1}^{d'} (z - \sigma'_i). \tag{6.36}$$

This relation shows that knowing the coefficients $I_{n'_1 j-1, k'_1 j}$, $j = 1, \ldots, d'$ from $I(z, w) = 0$ only determines the σ'_i up to some order. Thus knowledge of the ordering of the points $P_{\infty,i}$ is essential in our formulation. Now from

$$(a_{1,k'+1}, \ldots, a_{n-k',k'}) \in (\mathbb{C}^*)^{n-k'} \text{ and } (a_{n-k'+1,1}, \ldots, a_{n-d',k'-d'}) \in (\mathbb{C}^*)^{k'-d'},$$

we can construct

$$\tau'_i = \prod_{j=1}^{n'_1-k'_1} a_{i+(j-1)d', k'+i+(j-1)d'} \prod_{j=1}^{k'_1-1} a_{n-k'+i+(j-1)d', i+(j-1)d'}, \ 1 \le i \le d'. \tag{6.37}$$

We now define

$$a_{n-d'+i,i+(k_1'-1)d'} = \frac{\sigma_i'}{\tau_i'}, \quad 1 \le i \le d'. \tag{6.38}$$

Next, if we go through the argument in Proposition 5.4, it is easy to see that the Newton polygon of the given curve C is the same triangular region with vertices at the points $(0,n)$, $(n-k,0)$ and $(n+k',0)$. Moreover, $I(z,w)$ is non-degenerate with respect to its Newton polygon. (See the arguments in the proof of Theorem 5.6.)

STEP 2. On the other hand, it follows from the Riemann-Roch theorem and the regularity of D that

$$\begin{aligned} 1 \le &\dim \mathscr{L}\left(D + \sum_{i=1}^{j} P_{\infty,i} - \sum_{i=0}^{j-1} P_{0,i}\right) \\ \le &\dim \mathscr{L}\left(D + \sum_{i=1}^{j-1} P_{\infty,i} - \sum_{i=1}^{j-1} P_{0,i}\right) + 1 = 1 \\ \Rightarrow &\dim \mathscr{L}\left(D + \sum_{i=1}^{j} P_{\infty,i} - \sum_{i=0}^{j-1} P_{0,i}\right) = 1 \text{ for all } j. \end{aligned} \tag{6.39}$$

In the next few arguments, we will follow essentially what is in [**vMM**]. First of all, let f_j, determined up to a scalar, be the unique element in $\mathscr{L}(D+\sum_{i=1}^{j} P_{\infty,i} - \sum_{i=0}^{j-1} P_{0,i})$ and we normalize so that $zf_j = f_{j+n}$. As in previous calculations, we have

$$\begin{aligned} (z) &= n_1(P_{0,1}+\cdots+P_{0,d}) - n_1'(P_{\infty,1}+\cdots+P_{\infty,d'}), \\ (w) &= (n_1-k_1)(P_{0,1}+\cdots+P_{0,d}) - (n_1'+k_1')(P_{\infty,1}+\cdots+P_{\infty,d'}) \\ &\quad + \text{an effective divisor of degree } k+k' \end{aligned} \tag{6.40}$$

so that

$$\frac{w}{z} f_j \in \mathscr{L}\left(D + \sum_{i=1}^{k'+j} P_{\infty,i} - \sum_{i=0}^{j-k-1} P_{0,i}\right). \tag{6.41}$$

Since $k'+j > j-k$, by Lemma 5 in [**vMM**], it follows that $\frac{w}{z} f_j$ is a linear combination of f_i for $j-k \le i \le j+k'$. In other words, we have

$$\frac{w}{z} f_j = \sum_{i=-k}^{k'} c_{j,i+j} f_{i+j} \tag{6.42}$$

with $c_{j,j+k'} \ne 0$, $c_{j,j-k} \ne 0$. In this way, we obtain an infinite periodic band matrix $\mathbf{C} = (c_{ij})$ of period n modulo conjugation by a diagonal matrix of period n since f_j is only determined up to a multiplicative constant for each j. This is exactly what is obtained in [**vMM**].

STEP 3. We will now bring in the additional data as given by

(a) $(v_{n-d'+1}, \ldots, v_{n-1}, v_n = 1) \in (\mathbb{C}^*)^{d'}$,

(b) $(a_{1,k'+1}, \ldots, a_{n-k',n}) \in (\mathbb{C}^*)^{n-k'}$, and $(a_{n-k'+1,1}, \ldots, a_{n-d',k'-d'}) \in (\mathbb{C}^*)^{k'-d'}$

in Definition 6.4 (c) and (d).

For concreteness, for each j, we fix a $f_j^0 \in \mathscr{L}(D + \sum_{i=1}^{j} P_{\infty,i} - \sum_{i=0}^{j-1} P_{0,i})$. Then we have

$$\frac{w}{z} f_j^0 = \sum_{i=-k}^{k'} c_{j,i+j}^0 f_{i+j}^0 \tag{6.43}$$

with $c_{j,j+k'}^0 \neq 0$, $c_{j,j-k}^0 \neq 0$. Thus we have a unique periodic band matrix $\mathbf{C}^0 = (c_{ij}^0)$ corresponding to the choice of f_j^0, $j \in \mathbb{Z}$. Consequently (see (2.3) and Proposition 2.1), there exists $L^0 \in \mathcal{T}^{(k,k')}$ ($L^0(z) = z^{-1}L_{-1}^0 + L_0^0 + zL_1^0$), such that

$$\mathbf{C}^0 \delta_j = \sum_{i=0}^{n-1} (L^0(z))_{ij} \delta_i, \quad j = 0, \ldots, n-1 \tag{6.44}$$

and we will relabel the indices so that both the row index i and the column index j start from 1. Then

$$zL^0(z)\vec{f}^0 = w\vec{f}^0, \quad \vec{f}^0 = (f_1^0, \ldots, f_n^0)^T. \tag{6.45}$$

For the loop L^0 so obtained, it is not necessarily true that $(L_0^0)_{i,k'+i}$ is equal to $a_{i,k'+i}$, $i = 1, \ldots, n-k'$. Likewise, $(L_1^0)_{n-k'+i,i}$ is not necessary the same as $a_{n-k'+i,i}$, $i = 1, \ldots, k' - d'$. Let us set

$$\begin{aligned} &f_j = c_j f_j^0, \quad \widehat{v}_j^0 = \frac{f_j^0}{z}, \quad \widehat{v}_j = c_j \widehat{v}_j^0 \\ &c = \operatorname{diag}(c_1, \ldots, c_n). \end{aligned} \tag{6.46}$$

Then

$$\widehat{v} = c\widehat{v}^0. \tag{6.47}$$

Making use of this in (6.45), we obtain

$$zcL^0(z)c^{-1}\widehat{v} = w\widehat{v}. \tag{6.48}$$

Note that from (6.40) and $f_j^0 \in \mathscr{L}(D + \sum_{i=1}^{j} P_{\infty,i} - \sum_{i=0}^{j-1} P_{0,i})$, we have

$$(\widehat{v}_j^0) \geq -D + \sum_{i=j+1}^{n} P_{\infty,i} - \sum_{i=j}^{n-1} P_{0,i} \quad \text{for all } j. \tag{6.49}$$

Since $\operatorname{supp} D \cap \{P_{\infty,1}, \ldots, P_{\infty,d'}\} = \emptyset$, it follows from (6.49) that $\widehat{v}_{n-d'+i}^0(P_{\infty,i})$ exists and is nonzero, $i = 1, \ldots, d'$. Now we impose the condition

$$\begin{aligned} &(c_{n-d'+1}\widehat{v}_{n-d'+1}^0(P_{\infty,1}), \ldots, c_n\widehat{v}_n^0(P_{\infty,d'})) \\ =&(v_{n-d'+1}, \ldots, v_{n-1}, 1). \end{aligned} \tag{6.50}$$

This allows us to determine the values of $c_{n-d'+1}, \ldots, c_n$. Set $L(z) := cL^0(z)c^{-1} = z^{-1}L_{-1} + L_0 + zL_1$, we will determine the remaining entries in the diagonal matrix c by using the two vectors $(a_{1,k'+1}, \ldots, a_{n-k',n})$ and $(a_{n-k'+1,1}, \ldots, a_{n-d',k'-d'})$. We would like these vectors to be the vectors coming from the k'-th superdiagonal of L_0 and the $(n-k')$-th subdiagonal of L_1. Following the same idea in the direct problem, we partition our matrices into $d' \times d'$ blocks. As before, let

$$\gamma_1 = \operatorname{diag}(c_1, \ldots, c_{d'}), \ldots, \gamma_{n_1'} = \operatorname{diag}(c_{n-d'+1}, \ldots, c_n). \tag{6.51}$$

Taking the diagonal of the $d' \times d'$ blocks on the k_1'-th block superdiagonal of L_0 and L_0^0, we have

$$\begin{pmatrix} D_1 & & \\ & \ddots & \\ & & D_{n_1'-k_1'} \end{pmatrix}, \begin{pmatrix} E_1 & & \\ & \ddots & \\ & & E_{n_1'-k_1'} \end{pmatrix}, \tag{6.52}$$

where the diagonal blocks are now given by

$$\begin{aligned} D_1 &= \operatorname{diag}((L_0)_{1,k'+1}, \dots, (L_0)_{d',k'+d'}), \\ &\vdots \\ D_{n_1'-k_1'} &= \operatorname{diag}((L_0)_{n-k'-d'+1,n-d'+1}, \dots, (L_0)_{n-k',n}), \\ E_1 &= \operatorname{diag}((L_0^0)_{1,k'+1}, \dots, (L_0^0)_{d',k'+d'}), \\ &\vdots \\ E_{n_1'-k_1'} &= \operatorname{diag}(L_0^0)_{n-k'-d'+1,n-d'+1}, \dots, (L_0^0)_{n-k',n}). \end{aligned} \tag{6.53}$$

Note that here we are using the same notation as in (6.21) and the reader will see why it is convenient to do so later on (see (6.56)). Similarly, by taking the diagonal of the $d' \times d'$ blocks on the $(n_1' - k_1')$-th block subdiagonal of L_1 and L_1^0, we find (using the same notation as in (6.23))

$$\begin{pmatrix} D_{n_1'-k_1'+1} & & \\ & \ddots & \\ & & D_{n_1'} \end{pmatrix}, \begin{pmatrix} E_{n_1'-k_1'+1} & & \\ & \ddots & \\ & & E_{n_1'} \end{pmatrix}, \tag{6.54}$$

where the diagonal matrices are

$$\begin{aligned} D_{n_1'-k_1'+1} &= \operatorname{diag}((L_1)_{n-k'+1,1}, \dots, (L_1)_{n-k'+d',d'}), \\ &\vdots \\ D_{n_1'} &= \operatorname{diag}((L_1)_{n-d'+1,k'-d'+1}, \dots, (L_1)_{nk'}), \\ E_{n_1'-k_1'+1} &= \operatorname{diag}((L_1^0)_{n-k'+1,1}, \dots, (L^0)_1)_{n-k'+d',d'}), \\ &\vdots \\ E_{n_1'} &= \operatorname{diag}((L_1^0)_{n-d'+1,k'-d'+1}, \dots, (L_1^0)_{nk'}). \end{aligned} \tag{6.55}$$

Therefore, when we equate the diagonal of the $d' \times d'$ blocks on the k_1'-th block superdiagonal (resp. $(n_1' - k_1')$-th block subdiagonal) of the equation $L(z)c = cL^0(z)$, we obtain the equations

$$\gamma_i E_i = D_i \gamma_{\overline{\sigma}(i)}, \quad i = 1, \dots, n_1' \tag{6.56}$$

with the permutation in (6.27). Now we impose the conditions that

$$
\begin{aligned}
&((L_0)_{1,k'+1},\dots,(L_0)_{n-k',n}) = (a_{1,k'+1},\dots,a_{n-k',n}),\\
(6.57)\qquad &((L_1)_{n-k'+1,1},\dots,(L_1)_{n-d',k'-d'}) = (a_{n-k'+1,1},\dots,a_{n-d',k'-d'}).\\
&((L_1)_{n-d'+1,k'-d'+1},\dots,(L_1)_{nk'}) = (a_{n-d'+1,k'-d'+1},\dots,a_{nk'}),
\end{aligned}
$$

where we have used the quantities constructed in (6.38) (from Step 1) in the last line above. This determines the diagonal matrices $D_1, \dots, D_{n'_1}$. Since we know $E_1, \dots, E_{n'_1}$ and we know $\gamma_{n'_1}$ from (6.50), we can determine the other γ_i's, as in the section on the forward problem. Hence $L(z)$ is completely determined, and so is $\widehat{v}$.

STEP 4. In this step, we will show that the matrix loop $L(z)$ constructed above has the given set of admissible data as the associated extended algebro-geometric data. First of all, if $z \in \mathbb{CP}^1$ is not a branch point of C, then there are n points $P_1(z) = [1 : z : w_1(z)], \dots, P_n(z) = [1 : z : w_n(z)]$ on C lying over z. Define $\widehat{V}(z)$ as in (6.17), it follows from (6.48) that

$$
(6.58)\qquad zL(z)\widehat{V}(z) = \widehat{V}\Lambda(z), \quad \Lambda(z) = \operatorname{diag}(w_1(z),\dots,w_n(z)).
$$

Therefore $I(L,z,w) = I(z,w)$. On the other hand, since $f_n = z$, $zL(z)\bar{f} = w\bar{f}$ for $\bar{f} = (f_1, f_2, \dots, z)$ (this follows from (6.45), $\bar{f} = c\bar{f}^0$ and $L(z) = cL^0(z)c^{-1}$), and $f_j \in \mathscr{L}(D + \sum_{i=1}^{j} P_{\infty,i} - \sum_{i=0}^{j-1} P_{0,i})$), it follows that $D(L) = D$. Furthermore, by (6.47) and (6.50), we have $\widehat{v}_{n-d'+i}(P_{\infty,i}) = v_{n-d'+i}$, $i = 1, \dots, d'$. Finally, the conditions in (6.57) show that the numbers $a_{i,k'+i}$, $i = 1, \dots, n-k'$ are precisely the entries of L_0 on its k'-th superdiagonal, and that the numbers $a_{n-k'+i,i}$, $i = 1, \dots, k'-d'$ are the first $k'-d'$ entries of L_1 from its $(n-k')$-th subdiagonal.

STEP 5. In this final step, we will check that the matrix loop L constructed above is in $\mathcal{T}_*^{(k,k')}$. First of all, from part (a) of Definition 6.3, the σ_i's are nonzero and distinct, and the same holds true for the σ'_i. Therefore we have $L \in \mathcal{G}$. Since $I(L,z,w) = I(z,w)$ and we assume C has singular points only at P_0 and P_∞, it follows that if $\bar{I}_1(L,z)$, $\bar{I}_2(L,z)$ and $\bar{I}_3(L,z)$ are defined by $I(L,z,0) = z^{n-k}\bar{I}_1(L,z)$, $\frac{\partial I}{\partial z}(L,z,0) = z^{n-k-1}\bar{I}_2(L,z)$, $\frac{\partial I}{\partial w}(L,z,0) = z^{n-k}\bar{I}_3(L,z)$, and $R_{\alpha,\beta}(L)$ are defined by (5.26), then $R_{\alpha,\beta}(L) \neq 0$ for all α, β. Also we have $\Delta_A(I(L,\cdot,\cdot)) \neq 0$, where Δ_A is the A-resultant for A given by the set of lattice points in the triangular region with vertices at $(0,n)$, $(n-k,0)$ and $(n+k',0)$. Finally, the last condition in Definition 6.1 is clearly satisfied. This completes the checking. $\square$

CHAPTER 7

Solving the Equation via Factorization Problems

In this chapter, we will solve explicitly the Cauchy problem

$$\dot{L}(z) = [L(z), (\Pi_{\widetilde{\mathfrak{n}}_-} L)(z)], \quad L(z,0) = L^0(z) \tag{7.1}$$

where $L^0 \in \mathcal{T}_*^{(k,k')}$. We will solve this problem on its maximal interval of existence $[0,T)$. According to the group-theoretic setup, in order to do so, it suffices to solve for $g_+(\cdot,t) \in \widetilde{B}_+$, $g_-(\cdot,t) \in \widetilde{N}_-$ in the Riemann-Hilbert factorization problem

$$e^{tL^0(z)} = g_+(z,t)g_-(z,t)^{-1}, \quad z \in S^1. \tag{7.2}$$

THEOREM 7.1. *For $t \in [0,T)$, there exist unique holomorphic matrix-valued functions*

$$\begin{aligned} g_+(\cdot,t) &: \mathbb{CP}^1 \setminus \{\infty\} \longrightarrow GL(n,\mathbb{C}), \\ g_-(\cdot,t) &: \mathbb{CP}^1 \setminus \{0\} \longrightarrow GL(n,\mathbb{C}), \end{aligned} \tag{7.3}$$

which are smooth in $t \in [0,T)$, solve the factorization problem

$$e^{tL^0(z)} = g_+(z,t)g_-(z,t)^{-1}, \quad z \in \mathbb{CP}^1 \setminus \{0,\infty\} \tag{7.4}$$

and satisfy

$$g_+(\cdot,t)|_{S^1} \in \widetilde{B}_+, \quad g_-(\cdot,t)|_{S^1} \in \widetilde{N}_-. \tag{7.5}$$

Moreover, for initial data $L^0 \in \mathcal{T}_*^{(k,k')}$, $g_\pm(z,t)$ can be constructed by means of Riemann theta functions associated with the Riemann surface of the curve $\widetilde{C}(L^0)$ and the solution of the Cauchy problem (7.1) *is given by*

$$L(z,t) = g_\pm(z,t)^{-1}L^0(z)g_\pm(z,t). \tag{7.6}$$

By using the existence theorem of ODEs, we can establish the existence of the factors $g_\pm(\cdot,t)$ in the above theorem following the kind of arguments in [**DL**]. For this reason, we will omit details of the proof. In the rest of the section, we will show how to construct $g_\pm(z,t)$ via Riemann theta functions.

Denote by $\widetilde{C}(L^0)$ the normalization of the spectral curve $C(L^0)$ whose affine part is given by the equation $\det(zL^0(z)-wI)=0$. Let $\widehat{v}(P)$ satisfies the eigenvalue problem

$$z(P)L^0(z(P))\widehat{v}(P) = w(P)\widehat{v}(P), \ (e_n, \widehat{v}(P)) = 1 \tag{7.7}$$

for P on the curve away from the poles. If we let

$$\mu(P) = \frac{w}{z}(P) \tag{7.8}$$

and define

$$v_\pm(t,P) = g_\pm(z(P),t)^{-1}\widehat{v}(P), \tag{7.9}$$

then it follows from (7.4) and (7.7) that

$$\begin{aligned} &v_-(t,P) = e^{t\mu(P)}v_+(t,P), \quad z(P) \in \mathbb{CP}^1 \setminus \{0,\infty\}, \\ &L(z(P),t)v_\pm(t,P) = \mu(P)v_\pm(t,P). \end{aligned} \tag{7.10}$$

Hence the components of $v_\pm(t,P)$ satisfy scalar factorization problems (cf. [**DL**], [**RS**]).

In our next step, we will solve the scalar factorization problems. First, from the property of $g_-(z,t)$ and (5.50), we have

$$(v_-^j(t,\cdot)) \geq -D(L^0) \quad \text{on } \widetilde{C}(L^0) \setminus \{P_{0,1},\dots,P_{0,d}\}. \tag{7.11}$$

Similarly, we find that

$$(v_+^j(t,\cdot)) \geq -D(L^0) - \sum_{a=j}^{n-1} P_{0,a} \quad \text{on } \widetilde{C}(L^0) \setminus \{P_{\infty,1}\cdots,P_{\infty,d'}\}. \tag{7.12}$$

We will first solve the following scalar factorization problem

$$\begin{aligned} &\omega_-(t,P) = e^{t\mu(P)}\omega_+(t,P), \quad P \in \widetilde{C}(L^0) \setminus (\{P_{0,a}\}_{a=1}^d \cup \{P_{\infty,b}\}_{b=1}^{d'}), \\ &(\omega_+(t,\cdot)) \geq -D(L^0) \quad \text{on } \widetilde{C}(L^0) \setminus \{P_{\infty,1}\cdots,P_{\infty,d'}\}, \\ &(\omega_-(t,\cdot)) \geq -D(L^0) \quad \text{on } \widetilde{C}(L^0) \setminus \{P_{0,1}\cdots,P_{0,d}\}. \end{aligned} \tag{7.13}$$

To write down explicit solution of this problem, fix a canonical homology basis of $\widetilde{C}(L^0)$ and introduce the Riemann theta function, as in Section 6.1. Indeed, we will use the same notations from that subsection in what follows, with the understanding that the quantities are associated with $\widetilde{C}(L^0)$. We next introduce the unique abelian differential $d\phi_+$ of the second kind with vanishing a-periods that has poles only at $\{P_{\infty,1},\dots,P_{\infty,d'}\}$ and such that $d\phi_-(P) = (d\phi_+(P) + d\mu)(P)$ is regular on $\widetilde{C}(L^0)\setminus\{P_{0,1},\dots,P_{0,d}\}$ (for the existence of $d\phi_+$, the reader can consult p.47 and p. 67 of [**FK**]). Set

$$\omega_+^t(P) = \exp(t\phi_+(P)) \frac{\theta\left(e + \int_{D(L^0)}^{D_{g-1}+P}\omega + t\Phi_+\right)}{\theta\left(e + \int_{D(L^0)}^{D_{g-1}+P}\omega\right)}, \tag{7.14}$$

where $\phi_+(P) = \int_{P_{\infty,1}}^P d\phi_+$ and Φ_+ is the vector of b-peroids of $d\phi_+$, i.e.,

$$(\Phi_+)_j = \frac{1}{2\pi i}\int_{b_j} d\phi_+, j = 1,\dots,g = g(\widetilde{C}(L^0)). \tag{7.15}$$

It is easy to check that $\omega_+^t(P)$ is single-valued, and is meromorphic in $\widetilde{C}(L^0)\setminus\{P_{\infty,1},\dots,P_{\infty,d'}\}$ with $(\omega_+^t) \geq -D(L^0)$ there. Moreover, $e^{t\mu(P)}\omega_+^t(P)$ is meromorphic in $\widetilde{C}(L^0)\setminus\{P_{0,1},\dots,P_{0,d}\}$ with $(e^{t\mu}\omega_+^t) \geq -D(L^0)$ there. Consequently, ω_+^t and $\omega_-^t := e^{t\mu}\omega_+^t$ solve the factorization problem in (7.13).

Next, by comparing the factorization problems (7.10) and (7.13), we have

$$\omega_+^t(P)^{-1}v_+(t,P) = \omega_-^t(P)v_-(t,P), \quad P \in \widetilde{C}(L^0)\setminus(\{P_{0,a}\}_{a=1}^d \cup \{P_{\infty,b}\}_{b=1}^{d'}). \tag{7.16}$$

Consequently, the function defined by the relations

$$\widetilde{v}(t,P) \equiv \begin{cases} \omega_+^t(P)^{-1}v_+(t,P), & P \in \widetilde{C}(L^0)\setminus\{P_{\infty,1},\dots,P_{\infty,d'}\}, \\ \omega_-(P)^{-1}v_-(t,P), & P \in \widetilde{C}(L^0)\setminus\{P_{0,1},\dots,P_{0,d}\} \end{cases} \tag{7.17}$$

is meromorphic in $\widetilde{C}(L^0)$.

In the above discussion, it has been assumed that t is small throughout. In particular, it follows from this assumption that the function

$$\theta\left(e + \int_{D(L^0)}^{D_{g-1}+P} \omega + t\Phi_+\right) \tag{7.18}$$

which appeared in (7.14) has precisely g zeros. We will denote the corresponding divisor by $\widetilde{D}(t)$. Now it follows from (7.17), (7.14) and (7.12) that

$$(\widetilde{v}_j(t,\cdot)) \geq -\widetilde{D}(t) - \sum_{a=j}^{n-1} P_{0,a} \quad \text{on } \widetilde{C}(L^0) \setminus \{P_{\infty,1} \cdots, P_{\infty,d'}\}. \tag{7.19}$$

On the other hand, if we let $\widehat{v}^t(P)$ be the eigenvector of $L(z(P),t)$ with last component normalized to 1 and let $D(t)$ be its divisor of poles, i.e.,

$$(\widehat{v}_i^t) \geq -D(t) + \sum_{a=i+1}^{n} P_{\infty,a} - \sum_{a=i}^{n-1} P_{0,a} \tag{7.20}$$

for all i. We have the following result.

PROPOSITION 7.2. $D(t) = \widetilde{D}(t)$.

PROOF. From the definition of $\widetilde{v}(t,P)$ above and the second line of (7.10), we have

$$L(z(P),t)\widetilde{v}(t,P) = \mu(P)\widetilde{v}(t,P). \tag{7.21}$$

On the other hand, we also have

$$L(z(P),t)\widehat{v}^t(P) = \mu(P)\widehat{v}^t(P). \tag{7.22}$$

Therefore, $\widetilde{v}(t,P) = \widetilde{v}_n(t,P)\widehat{v}^t(P)$. Let $\widetilde{D}_0(t)$ be the divisor of zeros of $\widetilde{v}_n(t,P)$, then $(\widetilde{v}_n(t,\cdot)) = \widetilde{D}_0(t) - \widetilde{D}(t)$. Therefore, from the relation between $\widetilde{v}(t,P)$ and $\widehat{v}^t(P)$ above and (7.19), we find

$$\begin{aligned}(\widehat{v}_i^t) &= -(\widetilde{v}_n(t,\cdot)) + (\widetilde{v}_i(t,\cdot)) \\ &= -\widetilde{D}_0(t) + \widetilde{D}(t) + (\widetilde{v}_i(t,\cdot)) \\ &\geq -\widetilde{D}_0(t) - \sum_{a=i}^{n} P_{0,a}.\end{aligned} \tag{7.23}$$

From this relation, we conclude that $\widetilde{D}_0(t) \geq D(t)$ and as $\deg D(t) = \deg \widetilde{D}_0(t) = g$, we must have $\widetilde{D}_0(t) = D(t)$. But since $D(t)$ is a general divisor, so it follows from $(\widetilde{v}_n(t,\cdot)) = D(t) - \widetilde{D}(t)$ that $\widetilde{v}_n(t,P)$ is independent of P and this implies $D(t) = \widetilde{D}(t)$. □

From this proposition, it follows that we can solve for $\widetilde{v}_j(t,P)$ and hence $v_\pm^j(t,P)$ up to multiples for any j. Indeed, by using the prime form, we obtain

$$\widetilde{v}_j(t,P) = c_j(t)\,\frac{\prod_{a=j+1}^{n} E_e(P_{\infty,a},P)}{\prod_{b=j}^{n-1} E_e(P_{0,b},P)}\,\frac{\theta\left(e + \int_{\widetilde{D}(t)+\sum_{a=1}^{j} P_{\infty,a}}^{D_{g-1}+\sum_{b=0}^{j-1} P_{0,b}+P} \omega\right)}{\theta\left(e + \int_{\widetilde{D}(t)}^{D_{g-1}+P} \omega\right)}, \tag{7.24}$$

where $c_j(t)$ is to be determined. When this is combined with the formula for $\omega_+^t(P)$ in (7.14), we find

$$
\begin{aligned}
v_+^j(t,P) =& c_j(t)\exp(t\phi_+(P))\frac{\prod_{a=j+1}^{n} E_e(P_{\infty,a},P)}{\prod_{b=j}^{n-1} E_e(P_{0,b},P)} \frac{\theta\left(e+\int_{D(L^0)}^{D_{g-1}+P}\omega+t\Phi_+\right)}{\theta\left(e+\int_{D(L^0)}^{D_{g-1}+P}\omega\right)} \\
&\times\frac{\theta\left(e+\int_{\widetilde{D}(t)+\sum_{a=1}^{j}P_{\infty,a}}^{D_{g-1}+\sum_{b=0}^{j-1}P_{0,b}+P}\omega\right)}{\theta\left(e+\int_{\widetilde{D}(t)}^{D_{g-1}+P}\omega\right)},
\end{aligned}
\tag{7.25}
$$

with a similar formula for $v_-(t,P)$. Hence we can write

$$
v_\pm(t,P) = c(t)v_\pm^\theta(t,P) \tag{7.26}
$$

where $v_\pm^\theta(t,P)$ are known and $c(t)=\operatorname{diag}(c_1(t),\ldots,c_n(t))$ is to be determined.

We are now getting ready to construct $g_\pm(z,t)$ and in the process, we will determine $c(t)$. For given $z\in\mathbb{CP}^1$ which is not a branch point of the coordinate function $z(P)$, there exist n points $P_1(z)$, ..., $P_n(z)$ of the curve $\widetilde{C}(L^0)$ lying over z. Therefore we can define the matrices

$$
\begin{aligned}
V_\pm(z,t) &= (v_\pm(t,P_1(z))\cdots v_\pm(t,P_n(z))) \\
V_\pm^\theta(z,t) &= (v_\pm^\theta(t,P_1(z))\cdots v_\pm^\theta(t,P_n(z))) \\
\widehat{V}(z) &= (\widehat{v}(P_1(z))\cdots \widehat{v}(P_n(z)))
\end{aligned}
\tag{7.27}
$$

where $\widehat{v}(P)$ is determined by (6.13) with c_1, ..., c_n given by (6.16) and (6.26) starting with $\gamma_{n_1'}$. With these definitions, we have

$$
V_\pm(z,t) = g_\pm(z,t)^{-1}\widehat{V}(z) \tag{7.28}
$$

and so $g_\pm(z,t)$ are given by

$$
g_\pm(z,t) = \widehat{V}(z)V_\pm(z,t)^{-1} = \widehat{V}(z)V_\pm^\theta(z,t)^{-1}c(t)^{-1}. \tag{7.29}
$$

To determine $c(t)$, we use the fact that $g_-(\infty,t)$ exists and $g_-(\infty,t)\in N_-$. Consequently, $\lim_{z\to\infty}\widehat{V}(z)V_\pm^\theta(z,t)^{-1}$ exists and is an invertible lower triangular matrix. Therefore the diagonal entries of $c(t)$ are given by the formulas

$$
c_i(t) = (\lim_{z\to\infty}\widehat{V}(z)V_\pm^\theta(z,t)^{-1})_{ii}, \quad i=1,\ldots,n. \tag{7.30}
$$

This completes the construction of $g_\pm(z,t)$.

APPENDIX A

The Periodic Toda Lattice

For the convenience of the reader, we give in this appendix the reformulation of the periodic Toda lattice as a flow on infinite periodic band matrices with lower bandwidth and upper bandwidth both equal to 1 with 1's on the first superdiagonal. We will also write down the Lax pair formulation in terms of matrix loops $L(z) = L_1 z + L_0 + L_{-1}z^{-1}$ and $B(z)$, which are exactly the ones we find in equation (3) of [**BA**], except for the fact that L_0 and L_{-1} are now in $\mathfrak{gl}(n,\mathbb{R})$. Note that this formulation is different from the one used in [**AvM**], for example, because we are not using the realization of the periodic Toda lattice in terms of periodic Jacobi matrices here [**F**]. (This realization would involve Lie groups and Lie algebras different from the ones we are using here.)

To begin with, let us recall that the periodic Toda lattice is a system of n particles interacting with the Hamiltonian

$$H = \frac{1}{2}\sum_{j=1}^{n} p_j^2 + \sum_{j=1}^{n} e^{q_j - q_{j+1}} \tag{A.1}$$

and subject to periodic boundary conditions:

$$q_{j+n} = q_j, \quad p_{j+n} = p_j. \tag{A.2}$$

As such, the equations of motion of the n particles are given by

$$\dot{q}_j = p_j, \quad \dot{p}_j = e^{q_{j-1}-q_j} - e^{q_j - q_{j+1}}, \; j = 1, \ldots, n. \tag{A.3}$$

By analogy with the well-known Flaschka transformation [**F**], let

$$a_j = -p_j, \quad b_j = e^{q_j - q_{j+1}}, j = 1, \ldots, n, \tag{A.4}$$

then we can rewrite the equations of motion in the form

$$\dot{a}_j = b_j - b_{j-1}, \quad \dot{b}_j = b_j(a_{j+1} - a_j), \; j = 1, \ldots, n. \tag{A.5}$$

To express these equations using infinite periodic band matrices, let us first define what they are.

DEFINITION A.1. An infinite periodic band matrix of period n with lower bandwidth $k \geq 1$ and upper bandwidth $k' \geq 1$ is a two-sided matrix $\mathbf{A} = (a_{ij})_{i,j=-\infty}^{\infty}$ satisfying the following conditions:

(i) $a_{ij} = 0$ if $k < i - j$ or $j - i > k'$,
(ii) $a_{i+n,j+n} = a_{ij}$ for all i and j.

As an example, let us take $n = 4$, $k = 1$ and $k' = 2$, then

$$
\mathbf{A} = \begin{pmatrix}
\ddots & \ddots & \ddots & \ddots & \ddots & & & & & & & \\
0 & \beta_1 & \alpha_1 & \gamma_1 & \epsilon_1 & 0 & & & & & & \\
& 0 & \beta_2 & \alpha_2 & \gamma_2 & \epsilon_2 & 0 & & & & & \\
& & 0 & \beta_3 & \alpha_3 & \gamma_3 & \epsilon_3 & 0 & & & & \\
& & & 0 & \beta_4 & \alpha_4 & \gamma_4 & \epsilon_4 & 0 & & & \\
& & & & 0 & \beta_1 & \alpha_1 & \gamma_1 & \epsilon_1 & 0 & & \\
& & & & & 0 & \beta_2 & \alpha_2 & \gamma_2 & \epsilon_2 & 0 & \\
& & & & & & 0 & \beta_3 & \alpha_3 & \gamma_3 & \epsilon_3 & \ddots \\
& & & & & & & \ddots & \ddots & \ddots & \ddots & \ddots
\end{pmatrix}. \tag{A.6}
$$

Returning to the periodic Toda lattice, the infinite periodic band matrix that we need is of period n, with $k = k' = 1$ and has 1's on the first superdiagonal, given by the following:

$$
\mathbf{C}(a, b) = \begin{pmatrix}
\ddots & \ddots & & & & & & & & \\
\ddots & a_1 & 1 & & & & & & & \\
& b_1 & a_2 & 1 & & & & & & \\
& & b_2 & a_3 & 1 & & & & & \\
& & & \ddots & \ddots & \ddots & & & & \\
& & & & b_{n-1} & a_n & 1 & & & \\
& & & & & b_n & a_1 & 1 & & \\
& & & & & & b_1 & a_2 & 1 & \\
& & & & & & & b_2 & a_3 & 1 \\
& & & & & & & & \ddots & \ddots & \ddots
\end{pmatrix}. \tag{A.7}
$$

Therefore if $\mathbf{C}^+(a, b)$ is given by

$$
\mathbf{C}^+(a, b) = \begin{pmatrix}
\ddots & \ddots & & & & & & & \\
& a_1 & 1 & & & & & & \\
& & a_2 & 1 & & & & & \\
& & & \ddots & \ddots & & & & \\
& & & & a_n & 1 & & & \\
& & & & & a_1 & 1 & & \\
& & & & & & a_2 & 1 & \\
& & & & & & & a_3 & 1 \\
& & & & & & & & \ddots & \ddots
\end{pmatrix}, \tag{A.8}
$$

it is easy to check that the equations in (A.5) are equivalent to

$$
\dot{\mathbf{C}}(a, b) = [\mathbf{C}^+(a, b), \mathbf{C}(a, b)].
$$

Finally, the same equations in (A.5) can also be expressed in the form $\dot{L}(z) = [L(z), B(z)]$, where $L(z)$ and $B(z)$ are the following matrix loops, namely

$$L(z) = \begin{pmatrix} a_1 & 1 & 0 & \cdots & 0 & b_n z^{-1} \\ b_1 & a_2 & 1 & \ddots & & 0 \\ 0 & \ddots & \ddots & \ddots & \ddots & \vdots \\ \vdots & \ddots & \ddots & \ddots & \ddots & 0 \\ 0 & & \ddots & b_{n-2} & a_{n-1} & 1 \\ z & 0 & \cdots & 0 & b_{n-1} & a_n \end{pmatrix} = E_{n1} z + L_0 + L_{-1} z^{-1}, \tag{A.9}$$

and

$$B(z) = \begin{pmatrix} 0 & \cdots & \cdots & 0 & b_n z^{-1} \\ b_1 & 0 & & & 0 \\ 0 & \ddots & \ddots & & \vdots \\ \vdots & \ddots & b_{n-2} & \ddots & \vdots \\ 0 & \cdots & 0 & b_{n-1} & 0 \end{pmatrix} = \Pi_{\mathfrak{n}_-} L_0 + L_{-1} z^{-1}, \tag{A.10}$$

where $\Pi_{\mathfrak{n}_-} L_0$ is the strictly lower triangular part of L_0.

By comparing our derivation here with (1.5), (1.10)-(1.12), and Theorem A in Chapter 1, we thus see that what we are doing in this work has origins in the study of the periodic Toda lattice.

Bibliography

[A] M. Adler, *On a trace functional for formal pseudo differential operators and the symplectic structure of the Korteweg-de Vries type equations*, Invent. Math. **50** (1978/79), no. 3, 219–248, DOI 10.1007/BF01410079. MR520927

[AvM] M. Adler and P. van Moerbeke, *Completely integrable systems, Euclidean Lie algebras, and curves*, Adv. in Math. **38** (1980), no. 3, 267–317, DOI 10.1016/0001-8708(80)90007-9. MR597729

[AvMV] M. Adler, P. van Moerbeke, and P. Vanhaecke, *Algebraic integrability, Painlevé geometry and Lie algebras*, Ergebnisse der Mathematik und ihrer Grenzgebiete. 3. Folge. A Series of Modern Surveys in Mathematics [Results in Mathematics and Related Areas. 3rd Series. A Series of Modern Surveys in Mathematics], vol. 47, Springer-Verlag, Berlin, 2004, DOI 10.1007/978-3-662-05650-9. MR2095251

[Ap] A. I. Aptekarev, *Asymptotic properties of polynomials orthogonal on a system of contours, and periodic motions of Toda chains* (Russian), Mat. Sb. (N.S.) **125(167)** (1984), no. 2, 231–258. MR764479

[Ar] J. M. Arnold, *Soliton pulse-position modulation*, IEEE Proc.-J **140** (1993), 359-366.

[Aud] M. Audin, *Hamiltonian systems and their integrability*, SMF/AMS Texts and Monographs, vol. 15, American Mathematical Society, Providence, RI; Société Mathématique de France, Paris, 2008. Translated from the 2001 French original by Anna Pierrehumbert; Translation edited by Donald Babbitt. MR2440371

[B] H. F. Baker, *Examples of applications of Newton's polygon to the theory of singular points of algebraic functions*, Trans. Cambridge Philos. Soc. **15** (1893), 403-450.

[BA] K. Ben Abdeljelil, *The integrability of the periodic full Kostant-Toda lattice on a simple Lie algebra*, J. Lie Theory **21** (2011), no. 4, 929–960. MR2917699

[BK] E. Brieskorn and H. Knörrer, *Plane algebraic curves*, Birkhäuser Verlag, Basel, 1986. Translated from the German by John Stillwell, DOI 10.1007/978-3-0348-5097-1. MR886476

[BP] P. Beelen and R. Pellikaan, *The Newton polygon of plane curves with many rational points*, Des. Codes Cryptogr. **21** (2000), no. 1-3, 41–67, DOI 10.1023/A:1008323208670. Special issue dedicated to Dr. Jaap Seidel on the occasion of his 80th birthday (Oisterwijk, 1999). MR1801161

[Br] D. M. Bressoud, *Proofs and confirmations*, MAA Spectrum, Mathematical Association of America, Washington, DC; Cambridge University Press, Cambridge, 1999. The story of the alternating sign matrix conjecture. MR1718370

[CC] W. Castryck and F. Cools, *On the intrinsicness of the Newton polygon*, `arXiv:1304:4997`, 2013.

[DLNT] P. Deift, L. C. Li, T. Nanda, and C. Tomei, *The Toda flow on a generic orbit is integrable*, Comm. Pure Appl. Math. **39** (1986), no. 2, 183–232, DOI 10.1002/cpa.3160390203. MR820068

[DLT1] P. Deift, L. C. Li, and C. Tomei, *Matrix factorizations and integrable systems*, Comm. Pure Appl. Math. **42** (1989), no. 4, 443–521, DOI 10.1002/cpa.3160420405. MR990138

[DLT2] P. Deift, L.-C. Li, and C. Tomei, *Symplectic aspects of some eigenvalue algorithms*, Important developments in soliton theory, Springer Ser. Nonlinear Dynam., Springer, Berlin, 1993, pp. 511–536. MR1280489

[DL] P. A. Deift and L. C. Li, *Generalized affine Lie algebras and the solution of a class of flows associated with the QR eigenvalue algorithm*, Comm. Pure Appl. Math. **42** (1989), no. 7, 963–991, DOI 10.1002/cpa.3160420704. MR1008798

[DMN] B. A. Dubrovin, V. B. Matveev, and S. P. Novikov, *Nonlinear equations of Korteweg-de Vries type, finite-band linear operators and Abelian varieties* (Russian), Uspehi Mat. Nauk **31** (1976), no. 1(187), 55–136. MR427869

[EFS] N. M. Ercolani, H. Flaschka, and S. Singer, *The geometry of the full Kostant-Toda lattice*, Integrable systems (Luminy, 1991), Progr. Math., vol. 115, Birkhäuser Boston, Boston, MA, 1993, pp. 181–225. MR1279823

[F] H. Flaschka, *The Toda lattice. I. Existence of integrals*, Phys. Rev. B (3) **9** (1974), 1924–1925. MR408647

[FK] H. M. Farkas and I. Kra, *Riemann surfaces*, Graduate Texts in Mathematics, vol. 71, Springer-Verlag, New York-Berlin, 1980. MR583745

[GK] A. Givental and B. Kim, *Quantum cohomology of flag manifolds and Toda lattices*, Comm. Math. Phys. **168** (1995), no. 3, 609–641. MR1328256

[GZK] I. M. Gel′fand, A. V. Zelevinskiĭ, and M. M. Kapranov, *A-discriminants and Cayley-Koszul complexes* (Russian), Dokl. Akad. Nauk SSSR **307** (1989), no. 6, 1307–1311; English transl., Soviet Math. Dokl. **40** (1990), no. 1, 239–243. MR1020868

[GKZ] I. M. Gel′fand, M. M. Kapranov, and A. V. Zelevinsky, *Discriminants, resultants, and multidimensional determinants*, Mathematics: Theory & Applications, Birkhäuser Boston, Inc., Boston, MA, 1994, DOI 10.1007/978-0-8176-4771-1. MR1264417

[GS] M. I. Gekhtman and M. Z. Shapiro, *Noncommutative and commutative integrability of generic Toda flows in simple Lie algebras*, Comm. Pure Appl. Math. **52** (1999), no. 1, 53–84, DOI 10.1002/(SICI)1097-0312(199901)52:1<53::AID-CPA3>3.0.CO;2-5. MR1648421

[Gu] M. A. Guest, *Harmonic maps, loop groups, and integrable systems*, London Mathematical Society Student Texts, vol. 38, Cambridge University Press, Cambridge, 1997, DOI 10.1017/CBO9781139174848. MR1630443

[Kho] A. G. Hovanskiĭ, *Newton polyhedra, and the genus of complete intersections* (Russian), Funktsional. Anal. i Prilozhen. **12** (1978), no. 1, 51–61. MR487230

[Kir] F. Kirwan, *Complex algebraic curves*, London Mathematical Society Student Texts, vol. 23, Cambridge University Press, Cambridge, 1992, DOI 10.1017/CBO9780511623929. MR1159092

[K1] B. Kostant, *Quantization and unitary representations. I. Prequantization*, Lectures in Modern Analysis and Applications, III, Lecture Notes in Math., Vol. 170, Springer, Berlin-New York, 1970, pp. 87–208. MR294568

[K2] B. Kostant, *The solution to a generalized Toda lattice and representation theory*, Adv. in Math. **34** (1979), no. 3, 195–338, DOI 10.1016/0001-8708(79)90057-4. MR550790

[Kron] L. Kronecker, *Grundzüge einer arithmetischen Theorie der algebraische Grössen* (German), J. Reine Angew. Math. **92** (1882), 1–122, DOI 10.1515/crll.1882.92.1. MR1579896

[Kn] D. E. Knuth, *The art of computer programming. Volume 1: Fundamental algorithms*, 2nd ed., Addison-Wesley Series in Computer Science and Information Processing, Addison-Wesley Publishing Co., Reading, Mass.-London-Amsterdam, 1975. Exercise 37. MR0378456

[LN1] L.-C. Li and Z. Nie, *Integrability of the periodic Kostant-Toda flow on matrix loops of level* k (English, with English and French summaries), C. R. Math. Acad. Sci. Paris **353** (2015), no. 4, 363–367, DOI 10.1016/j.crma.2015.01.006. MR3319135

[LN2] L.-C. Li and Z. Nie, *On the Liouville integrability of the periodic Kostant-Toda flow on matrix loops of level* k, Comm. Math. Phys. **352** (2017), no. 3, 1153–1203, DOI 10.1007/s00220-016-2768-7. MR3631402

[M] S. V. Manakov, *Complete integrability and stochastization of discrete dynamical systems* (Russian, with English summary), Ž. Èksper. Teoret. Fiz. **67** (1974), no. 2, 543–555; English transl., Soviet Physics JETP **40** (1974), no. 2, 269–274 (1975). MR389107

[MOP] A. V. Mikhaĭlov, M. A. Olshanetsky, and A. M. Perelomov, *Two-dimensional generalized Toda lattice*, Comm. Math. Phys. **79** (1981), no. 4, 473–488. MR623963

[MSC] V. Muto, A. Scott, and P. Christiansen, *A Toda lattice model for DNA: thermally generated solitons*, Physica D **44** (1990), 75-91.

[MvM] H. P. McKean and P. van Moerbeke, *The spectrum of Hill's equation*, Invent. Math. **30** (1975), no. 3, 217–274, DOI 10.1007/BF01425567. MR397076

[Mum] D. Mumford, *Tata lectures on theta. I*, Progress in Mathematics, vol. 28, Birkhäuser Boston, Inc., Boston, MA, 1983. With the assistance of C. Musili, M. Nori, E. Previato and M. Stillman, DOI 10.1007/978-1-4899-2843-6. MR688651

[N] M. Noether, *Rationale Ausführung der Operationen in der Theorie der algebraischen Functionen* (German), Math. Ann. **23** (1884), no. 3, 311–358, DOI 10.1007/BF01446394. MR1510258

[O] A. M. Ostrowski, *On Kronecker's elimination theory*, J. Reine Angew. Math. **296** (1977), 37–57, DOI 10.1515/crll.1977.296.37. MR498524

[OT1] D. Olive and N. Turok, *The symmetries of Dynkin diagrams and the reduction of Toda field equations*, Nuclear Phys. B **215** (1983), no. 4, 470–494, DOI 10.1016/0550-3213(83)90256-0. MR699743

[OT2] D. I. Olive and N. Turok, *Algebraic structure of Toda systems*, Nuclear Phys. B **220** (1983), no. 4, 491–507, DOI 10.1016/0550-3213(83)90504-7. MR717191

[P] O. Perron, *Algebra I*, De Gruyter, Berlin 1927.

[RS] A. G. Reyman and M. A. Semenov-Tian-Shansky, *Reduction of Hamiltonian systems, affine Lie algebras and Lax equations. II*, Invent. Math. **63** (1981), no. 3, 423–432, DOI 10.1007/BF01389063. MR620678

[STS] M. A. Semenov-Tyan-Shanskiĭ, *What a classical r-matrix is* (Russian), Funktsional. Anal. i Prilozhen. **17** (1983), no. 4, 17–33. MR725413

[Sy] W. W. Symes, *The QR algorithm and scattering for the finite nonperiodic Toda lattice*, Phys. D **4** (1981/82), no. 2, 275–280, DOI 10.1016/0167-2789(82)90069-0. MR653781

[T] M. Toda, *Vibration of a chain with nonlinear interaction*, J. Phys. Soc. Japan **22** (1967), 431-436.

[UT] K. Ueno and K. Takasaki, *Toda lattice hierarchy. I*, Proc. Japan Acad. Ser. A Math. Sci. **59** (1983), no. 5, 167–170. MR718796

[vM] P. van Moerbeke, *The spectrum of Jacobi matrices*, Invent. Math. **37** (1976), no. 1, 45–81, DOI 10.1007/BF01418827. MR650253

[vMM] P. van Moerbeke and D. Mumford, *The spectrum of difference operators and algebraic curves*, Acta Math. **143** (1979), no. 1-2, 93–154, DOI 10.1007/BF02392090. MR533894

[W] R. J. Walker, *Algebraic curves*, Springer-Verlag, New York-Heidelberg, 1978. Reprint of the 1950 edition. MR513824

[Wat] D. S. Watkins, *Isospectral flows*, SIAM Rev. **26** (1984), no. 3, 379–391, DOI 10.1137/1026075. MR750456

[Wi] G. Wilson, *The modified Lax and two-dimensional Toda lattice equations associated with simple Lie algebras*, Ergodic Theory Dynam. Systems **1** (1981), no. 3, 361–380 (1982), DOI 10.1017/s0143385700001292. MR662474

[ZMR] V. E. Zakharov, S. L. Musher, and A. M. Rubenchik, *Weak Langmuir turbulence of an isothermal plasma*, Soviet Physics JETP **42** (1975) 80-86.

Editorial Information

To be published in the *Memoirs*, a paper must be correct, new, nontrivial, and significant. Further, it must be well written and of interest to a substantial number of mathematicians. Piecemeal results, such as an inconclusive step toward an unproved major theorem or a minor variation on a known result, are in general not acceptable for publication.

Papers appearing in *Memoirs* are generally at least 80 and not more than 200 published pages in length. Papers less than 80 or more than 200 published pages require the approval of the Managing Editor of the Transactions/Memoirs Editorial Board. Published pages are the same size as those generated in the style files provided for $\mathcal{A}_{\mathcal{M}}\mathcal{S}$-LaTeX.

Information on the backlog for this journal can be found on the AMS website starting from `http://www.ams.org/memo`.

A Consent to Publish is required before we can begin processing your paper. After a paper is accepted for publication, the Providence office will send a Consent to Publish and Copyright Agreement to all authors of the paper. By submitting a paper to the *Memoirs*, authors certify that the results have not been submitted to nor are they under consideration for publication by another journal, conference proceedings, or similar publication.

Information for Authors

Memoirs is an author-prepared publication. Once formatted for print and on-line publication, articles will be published as is with the addition of AMS-prepared frontmatter and backmatter. Articles are not copyedited; however, confirmation copy will be sent to the authors.

Initial submission. The AMS uses Centralized Manuscript Processing for initial submissions. Authors should submit a PDF file using the Initial Manuscript Submission form found at `www.ams.org/submission/memo`, or send one copy of the manuscript to the following address: Centralized Manuscript Processing, MEMOIRS OF THE AMS, 201 Charles Street, Providence, RI 02904-2294 USA. If a paper copy is being forwarded to the AMS, indicate that it is for *Memoirs* and include the name of the corresponding author, contact information such as email address or mailing address, and the name of an appropriate Editor to review the paper (see the list of Editors below).

The paper must contain a *descriptive title* and an *abstract* that summarizes the article in language suitable for workers in the general field (algebra, analysis, etc.). The *descriptive title* should be short, but informative; useless or vague phrases such as "some remarks about" or "concerning" should be avoided. The *abstract* should be at least one complete sentence, and at most 300 words. Included with the footnotes to the paper should be the 2020 *Mathematics Subject Classification* representing the primary and secondary subjects of the article. The classifications are accessible from `www.ams.org/msc/`. The Mathematics Subject Classification footnote may be followed by a list of *key words and phrases* describing the subject matter of the article and taken from it. Journal abbreviations used in bibliographies are listed in the latest *Mathematical Reviews* annual index. The series abbreviations are also accessible from `www.ams.org/msnhtml/serials.pdf`. To help in preparing and verifying references, the AMS offers MR Lookup, a Reference Tool for Linking, at `www.ams.org/mrlookup/`.

Electronically prepared manuscripts. The AMS encourages electronically prepared manuscripts, with a strong preference for $\mathcal{A}_{\mathcal{M}}\mathcal{S}$-LaTeX. To this end, the Society has prepared $\mathcal{A}_{\mathcal{M}}\mathcal{S}$-LaTeX author packages for each AMS publication. Author packages include instructions for preparing electronic manuscripts, samples, and a style file that generates the particular design specifications of that publication series.

Authors may retrieve an author package for *Memoirs of the AMS* from `www.ams.org/journals/memo/memoauthorpac.html`. The *AMS Author Handbook* is available in PDF format from the author package link. The author package can also be obtained free of charge by sending email to `tech-support@ams.org` or from the Publication Division,

American Mathematical Society, 201 Charles St., Providence, RI 02904-2294, USA. When requesting an author package, please specify the publication in which your paper will appear. Please be sure to include your complete mailing address.

After acceptance. The source files for the final version of the electronic manuscript should be sent to the Providence office immediately after the paper has been accepted for publication. The author should also submit a PDF of the final version of the paper to the editor, who will forward a copy to the Providence office.

Accepted electronically prepared files can be submitted via the web at `www.ams.org/submit-book-journal/`, sent via FTP, or sent on CD to the Electronic Prepress Department, American Mathematical Society, 201 Charles Street, Providence, RI 02904-2294 USA. TEX source files and graphic files can be transferred over the Internet by FTP to the Internet node `ftp.ams.org` (130.44.1.100). When sending a manuscript electronically via CD, please be sure to include a message indicating that the paper is for the *Memoirs*.

Electronic graphics. Comprehensive instructions on preparing graphics are available at `www.ams.org/authors/journals.html`. A few of the major requirements are given here.

Submit files for graphics as EPS (Encapsulated PostScript) files. This includes graphics originated via a graphics application as well as scanned photographs or other computer-generated images. If this is not possible, TIFF files are acceptable as long as they can be opened in Adobe Photoshop or Illustrator.

Authors using graphics packages for the creation of electronic art should also avoid the use of any lines thinner than 0.5 points in width. Many graphics packages allow the user to specify a "hairline" for a very thin line. Hairlines often look acceptable when proofed on a typical laser printer. However, when produced on a high-resolution laser imagesetter, hairlines become nearly invisible and will be lost entirely in the final printing process.

Screens should be set to values between 15% and 85%. Screens which fall outside of this range are too light or too dark to print correctly. Variations of screens within a graphic should be no less than 10%.

Any graphics created in color will be rendered in grayscale for the printed version unless color printing is authorized by the Managing Editor and the Publisher. In general, color graphics will appear in color in the online version.

Inquiries. Any inquiries concerning a paper that has been accepted for publication should be sent to `memo-query@ams.org` or directly to the Electronic Prepress Department, American Mathematical Society, 201 Charles St., Providence, RI 02904-2294 USA.

SELECTED PUBLISHED TITLES IN THIS SERIES

1565 **Evan M. O'Dorney,** Reflection Theorems of Ohno-Nakagawa Type for Quartic Rings and Pairs of n-Ary Quadratic Forms, 2025

1564 **Alexander S. Kleshchev and Michael Livesey,** RoCK Blocks for Double Covers of Symmetric Groups and Quiver Hecke Superalgebras, 2025

1563 **Jacques Darné, Martin Palmer, and Arthur Soulié,** When the Lower Central Series Stops: A Comprehensive Study for Braid Groups and Their Relatives, 2025

1562 **Paolo Piazza, Thomas Schick, and Vito Felice Zenobi,** Mapping Analytic Surgery to Homology, Higher Rho Numbers and Metrics of Positive Scalar Curvature, 2025

1561 **Bahar Acu, Orsola Capovilla-Searle, Agnès Gadbled, Aleksandra Marinković, Emmy Murphy, Laura Starkston, and Angela Wu,** Weinstein Handlebodies for Complements of Smoothed Toric Divisors, 2025

1560 **Pranav V. Chakravarthy and Martin Pinsonnault,** Centralizers of Hamiltonian Circle Actions on Rational Ruled Surfaces, 2025

1559 **Nicholas M. Katz, Antonio Rojas-León, and Pham Huu Tiep,** Rigid Local Systems and Sporadic Simple Groups, 2025

1558 **Sergio Albeverio and Seiichiro Kusuoka,** Construction of a Non-Gaussian and Rotation-Invariant Φ^4-Measure and Associated Flow on $\mathbb{R}^3$ Through Stochastic Quantization, 2025

1557 **Akishi Ikeda and Yu Qiu,** q-Stability Conditions via q-Quadratic Differentials for Calabi-Yau-$\mathbb{X}$ Categories, 2025

1556 **Sumiya Baasandorj and Sun-Sig Byun,** Regularity for Orlicz Phase Problems, 2025

1555 **Damien Calaque, Rune Haugseng, and Claudia Scheimbauer,** The AKSZ Construction in Derived Algebraic Geometry as an Extended Topological Field Theory, 2025

1554 **Matteo Bonforte, Jean Dolbeault, Bruno Nazaret, and Nikita Simonov,** Stability in Gagliardo–Nirenberg–Sobolev Inequalities: Flows, Regularity and the Entropy Method, 2025

1553 **Andreas Greven, Frank den Hollander, and Margriet Oomen,** Spatial Populations with Seed-Bank: Renormalisation on the Hierarchical Group, 2025

1552 **Jason Atnip, Hiroki Sumi, and Mariusz Urbański,** The Dynamics and Geometry of Semi-Hyperbolic Rational Semigroups, 2025

1551 **Ronan J. Conlon and Alix Deruelle,** Steady Gradient Kähler-Ricci Solitons on Crepant Resolutions of Calabi-Yau Cones, 2025

1550 **Jussi Behrndt, Fritz Gesztesy, and Marius Mitrea,** Sharp Boundary Trace Theory and Schrödinger Operators on Bounded Lipschitz Domains, 2025

1549 **Valentina Grazian and Chris Parker,** Saturated Fusion Systems on p-Groups of Maximal Class, 2025

1548 **Rui Loja Fernandes and Ivan Struchiner,** The Global Solutions to a Cartan's Realization Problem, 2025

1547 **Satoshi Kondo and Seidai Yasuda,** Arithmetic Quotients of the Bruhat-Tits Building for Projective General Linear Groups in Positive Characteristic, 2025

1546 **Filippo Mazzoli, Andrea Seppi, and Andrea Tamburelli,** Para-HyperKähler Geometry of the Deformation Space of Maximal Globally Hyperbolic Anti-de Sitter Three-Manifolds, 2025

1545 **Tobias Lenz,** G-Global Homotopy Theory and Algebraic K-Theory, 2025

1544 **James Maynard,** Primes in Arithmetic Progressions to Large Moduli III: Uniform Residue Classes, 2025

1543 **James Maynard,** Primes in Arithmetic Progressions to Large Moduli II: Well-Factorable Estimates, 2025

Printed and bound by CPI Group (UK) Ltd, Croydon, CR0 4YY
09/11/2026
14999542-0002